Vanity, Vitality, and Virility

The science behind the products you love to buy

John Emsley won the Science Book prize in 1995 for his *Consumer's Good Chemical Guide*, and followed this with a series of popular science books: *Molecules at an Exhibition, Was it Something You Ate?* (co-authored with Peter Fell), *The Shocking History of Phosphorus, Nature's Building Blocks*, and *The Elements of Murder*, all of which have been translated worldwide.

John spent twenty years as a researcher and lecturer in chemistry at London University before becoming a freelance popular-science writer and a Science Writer in Residence, first at Imperial College London and then in the Chemistry Department of the University of Cambridge. In 2003 he was awarded the German Chemical Society's Writer's Award.

Praise for other books by John Emsley:

'A delightful potion of chemical erudition, forgotten science history and ghastly murder schemes. Along the way, the bodies pile up as Emsley relates spectacular case histories of poisonings, accidental and criminal . . .'
The New York Times on *The Elements of Murder*

'. . . fascinating, wide-ranging and, let's not mince words, macabre new history of poison . . . a truly guilty pleasure.'
The Wall Street Journal on *The Elements of Murder*

'What for many might be a dry and dusty collection of facts has been turned into an amusing and finely crafted set of mini-biographies . . . This is a fine, amusing and quirky book . . .'
Nature on *Nature's Building Blocks*

Vanity, Vitality, and Virility

The science behind the products
you love to buy

JOHN EMSLEY

OXFORD
UNIVERSITY PRESS

OXFORD
UNIVERSITY PRESS

Great Clarendon Street, Oxford OX2 6DP

Oxford University Press is a department of the University of Oxford.
It furthers the University's objective of excellence in research, scholarship,
and education by publishing worldwide in

Oxford New York

Auckland Cape Town Dar es Salaam Hong Kong Karachi
Kuala Lumpur Madrid Melbourne Mexico City Nairobi
New Delhi Shanghai Taipei Toronto

With offices in

Argentina Austria Brazil Chile Czech Republic France Greece
Guatemala Hungary Italy Japan Poland Portugal Singapore
South Korea Switzerland Thailand Turkey Ukraine Vietnam

Oxford is a registered trade mark of Oxford University Press
in the UK and in certain other countries

Published in the United States
by Oxford University Press Inc., New York

British Library Cataloguing in Publication Data

Data available

Library of Congress Cataloging in Publication Data

Data available

Designed and typeset by Pete Russell, Faringdon, Oxon.
Printed in Great Britain by
Clays Limited, St Ives plc

ISBN 0-19-280673-4   978-0-19-280673-4

1 3 5 7 9 10 8 6 4 2

The following trade names are referred to in this book: Absormex, Actal, Adams Chewing Gum, Aida,
Aludrox, AlzeimAlert, Ambre Solaire, Antarox BL-225, Benecol, BioAstin, Black and White Bleaching Cream,
Boots No.7, Cardilate, Caverject, Celaxa, Cialis, Cipramil, Clean Shower, Coro-Nitro, Deponit, Dettol,
Dubble Bubble, Elantan, Eldopaque, Favertin, FloraPro.active, Germinol, Hazlone, Huggies, Isoket, Isordil,
Jeyes Fluid, Kraton, Levitra, Lexan, Librium, LifeGem, Listerine, Lustral, Luvox, Lysol, Man-Tan, Melanotan,
Mogadon, Monit, Mycardol, Mystery Gum, Nature Boy & Girl, NatureWorks, Neo-Heliopan, Nivea, Oil of Olay,
Oxtone, Pampers, Parsol, Paxil, Pentothal, pHiso-Hex, Prozac, Quick Tan, Retin-A, Retinova, Savlon, SeaAcid,
Seconal, Sephiran, Seroxat, Synergie Wrinkle Lift, Technicolor, Tecquinol, Tofranil, Très BN, Valium,
Vasomax, Viagra, Vitanex, Westsal, Winter Tan, Wrigley's Spearmint Gum, Xantar, X-cite, and Zoloft.
The author and Oxford University Press acknowledge the rights of the owners of these marks.

Contents

Acknowledgements

DURING the writing of *Vanity, Vitality, and Virility*, I prevailed upon friends, colleagues, and acquaintances, whom I knew to be knowledgeable about the various topics covered, to read parts of it and give me their comments. Their help has been invaluable and I should like to express my heartfelt thanks to them.

Those who helped me get Chapter 1 right were popular-science author Philip Ball, consultant John Woodruff, and John Ballington, who is the Corporate & Consumer Affairs Director of Lever Fabergé. John also enlisted the help of other members of that company, notably Drs Graham Catton, Martin Jones, and Maria Labedzka, who made useful comments on the chapter's contents. Dr Tom Coultate, author of *Food: The Chemistry of its Components*, and Mrs Helen Glyn-Davies, state-registered dietitian of the Luton and Dunstable Hospital, gave useful advice on Chapter 2. Science writer, and author of the best-selling *Oxygen*, Dr Nick Lane, of Maida Vale, London, checked Chapters 3 and 5, and so did his wife, Ana, who is a medical doctor. Dr Dave Alker of Pfizer was consulted with regard to the Viagra section, while Dr Tom Welton of Imperial College, London, verified other parts of Chapter 3, and Dr Margaret Rayman of the Centre for Nutrition and Food Safety at Surrey University, Guildford, looked at the section on selenium and provided some up-to-the-minute information on the role of this element in human health. Angela Asieba, Diamond Information Manager of De Beers, London, scrutinized the section in Chapter 3 on diamonds. Dr Claire La-Rosa of Lever Fabergé was most helpful with Chapter 4, in particular with respect to its microbiological section. Dr Friedrich Sosna of Degussa AG, Germany, advised on anti-microbial polymers. The disposable nappy section of Chapter 6 was read by Dr Jac Lemmen, who works for Kimberly-Clark. Jo Hartop of Wrigley, Plymouth, and Chris Perille of Wrigley, Chicago, provided information on chewing gum. The whispering asphalt section was read by people who work for Kraton Polymers: Roger Morgan, who is Global Director based in the United Kingdom, Stephen Evans, who is the Global Business Manager and based in London, and Willem Vonk, Director of Kraton Polymers Holding BV, who is based in Amsterdam. Thanks also to Michael Berg for bringing to my attention the news about Viagra chewing gum. As for the Postscript and Glossary, I must take full responsibility.

A few close friends agreed to the imposition of reading the complete text, notably Walter Saxon of New York, who came up with extra information regarding hydrogen peroxide and the early photography of Niépce, and Professor Steve and Mrs Rose Ley of Cambridge University, who also made several positive suggestions.

Special thanks to Dr Michael Rodgers, my editor at Oxford University Press, who encouraged me to write the book and has seen it through its various stages, including negotiations with my agent Patrick Walsh of Conville & Walsh, of Soho, London, to whom I am also very grateful.

Finally, I must thank my wife, Joan, and other members of my family for their forbearance in listening to me go on about my favourite subject: the much-misunderstood role of chemistry and chemicals in everyday life.

Introduction

UNCERTAINTY is the hardest thing to live with, and the complexity of modern living leads to more and more uncertainty, especially about things that affect our lives and over which we feel we have only limited control. Will I get a job? Will I lose my job? Will someone be promoted over me? Can I pay off my debts? Will I be robbed? What if my partner finds someone else? Am I likely to get cancer? Are my children safe? Will a loved one die? These are not issues that *Vanity, Vitality, and Virility* can address, but there are other uncertainties that may have troubled you, and these I hope to tackle. These are the uncertainties about many of the products we use every day, such as cosmetics and household chemicals, or those we may occasionally encounter, such as treatments for sexual or mental dysfunctions, or those which we consume but are worried about, such as the various kinds of fat. *Vanity, Vitality, and Virility* was not written merely to set your mind at rest over issues that surround these various chemicals, but as a book to be enjoyed by those who simply want to know more about some of the chemicals that are now an intrinsic part of our lives.

Many people now regard anything described as 'chemical' as worrying, and at the same time regard natural materials as intrinsically preferable because they are risk free. As a chemist I find this comparison to be illogical and yet sadly prevalent. My aim in *Vanity, Vitality, and Virility* is not to prove that chemicals are safe and natural materials are a threat; indeed, many chemicals are highly dangerous and most natural things are perfectly safe. What I would like to show you is that many of the things that we take for granted in our lives are the products of chemistry, that they actually make life better, and that we are wrong to be suspicious of them.

The days when the chemical industry was careless and uncaring are long gone, and yet even in its early years it was responsible for rescuing millions of people from lives of dirt, scarcity, disease, drabness, and poverty, and what was achieved in the Western world fifty years ago can now be shared by all. The challenge is to spread the benefits without damaging the environment, by using only renewable resources, and there is no reason why this cannot be done. Indeed, by the end of this century I can envisage a chemical industry that is converting waste materials to useful products in chemical plants that occupy only a fraction of the space currently needed, perhaps even sited underground, and with no discharges to the environment.

Meanwhile, we should be alert to what can, and what cannot, be achieved by chemistry, and *Vanity, Vitality, and Virility* highlights its roles and limitations in such areas as cosmetics, food, sex, hygiene, and mental health. There is also a chapter, 'Polymers in Disguise', which shows how it impinges in areas that you might not expect—or even approve of. Finally, there is a Postscript in which I will set out the dangers of the current widespread opposition to all things 'chemical', and try to convince the 'chemisceptics' and the 'chemiphobics' that their suspicions and fears may be unfounded.

I have tried to make *Vanity, Vitality, and Virility* reader-friendly, and to do so I have avoided as much technical jargon as possible. My object has been to ensure that any educated person will be able to understand what I have written, whether they have a science background or not. I have stuck to common names for chemicals, and I hope that those who have a chemistry degree, and especially those who teach chemistry, will understand my reasons for doing so and forgive me for not using technical names. I have also avoided technical explanations in the text but the reader who might wish for more information can consult the Glossary, and such topics are shown in **bold**.

I make no apology for including a few chemical formulae in this book, and for occasionally using them in place of the written word. I accept that some readers may find this annoying, especially if it is several years since they studied chemistry at school. I must confess to an ulterior motive in using formulae, in that I would like people to become familiar with this kind of shorthand so that they are not fazed when they meet such formulae out of context. Although most people know that oxygen is O_2, water is H_2O, NaCl is salt, and carbon dioxide is CO_2, this should not be the limit of their knowledge. It is more than likely that their parents and even grandparents knew many more and would have immediately recognised H_2S as hydrogen sulfide, H_2SO_4 as sulfuric acid, NaOH as sodium hydroxide (although they might have called it caustic soda), and Pb as lead. No doubt there are some readers who are scientists, and feel it should not be part of this book to try to *teach* chemistry but, in an age when the barrier to understanding chemistry seems at times to be almost insurmountable, I am willing to try, and I hope they will bear with me.

So welcome to *Vanity, Vitality, and Virility*. In this selection of everyday chemicals there will be some that you currently use, some that you worry might be harmful, some that you hope you will never need, some that are there to protect you and your family from disease, some in which you might find solace when times get hard, and some that you just did not realize were from the chemical industry at all. And once you have read *Vanity, Vitality, and Virility* I hope you will find that I have removed a little uncertainly from your life and convinced you that the messages you have been getting from the 'chemiphobics', and even from some advertisers, are far from being the truth, the whole truth, and nothing but the truth.

CHAPTER ONE

Vanity:
No More Wrinkles!

FEW OF US have the kind of perfect body that we see in advertisements, magazines, and films, and although we know that such perfection is more the work of a graphics department or make-up artist than a gift of Nature, we still seek to emulate it. In real life, we have to make the best of what our parents endowed us with, and then it is to chemistry that we turn for help. In this chapter we shall look at some chemicals that enhance our attractiveness, protecting our skin and disguising its defects.

Skin is tough, indeed so tough that relatively few natural substances can penetrate it. Skin is made up of several layers: the outside layer is the epidermis and this is made up of a top layer called the stratum corneum, and a deeper layer called the stratum basale, which is where skin cells form.[1] Cells are pushed upwards from the stratum basale until they are the dead cells of the stratum corneum, there to be washed off or shed as dust. The life of a skin cell from cradle to grave takes about a month and, as our skin begins to show signs of age, we must wage a constant battle to keep it looking young. With the aid of chemistry we can ensure that some of the ravages of time are reduced, or at least disguised.

There are many chemicals used in cosmetics and other beauty products. The claims made for them may bring only a wry smile to your face because they seem to promise too much. Can they really erase wrinkles and make us look young again? The answer is no, but a qualified no, because the ingredients now used in cosmetics can reduce the appearance of wrinkles, even if they don't remove them completely. They might

---

[1] Below the epidermis is the dermis, which consists of a variety of cells and structures, including hair follicles, sweat glands, nerve endings, etc. It is changes in the dermis that are mainly responsible for skin ageing.

even be able to do that in twenty years' time, and indeed real advances towards this goal have been made in the past. Over the years, chemistry has transformed skin care and cosmetics in some surprising ways, such as by making lipstick almost indelible, and giving cosmetics a silk-like feel on the skin. More importantly, chemicals can protect the skin against damage by the sun, or they can also give you an instant tan, and even make dark skin appear lighter.

Among the molecules that perform these minor miracles are the alpha-hydroxy acids (AHAs), ricinoleic acid, eosin, dihydroxyactetone (DHA), hydroquinone, ethylhexyl methoxycinnamate, titanium dioxide, boron nitride, and liposomes. These are the ingredients that help us to fight off the ravages of time.

Lipstick

Lipstick is the sexiest cosmetic, and soon after it was invented it became an indispensable item for most women, despite the many drawbacks of the early versions, which broke easily, were not indelible, and turned rancid. Today, it comes without these defects, and in a multitude of colours and textures.

The Technicolor films of the 1930s are probably the reason why bright-red lips became an acceptable part of female beauty—for centuries it had been assumed that they were to be seen only on prostitutes and actresses. The Technicolor process tended to make lips seem artificially red but, nevertheless, Hollywood films were setting standards that many wanted to emulate, and the best way of painting one's lips red was with a lipstick. This created a demand for a product that had been invented twenty years earlier, and the popularity of lipstick has continued to this day.

Indeed, lipstick's image was bolstered by popular songs such as Holt Marvell's 1935 romantic hit, *These foolish things remind me of you*, which began with the line 'A cigarette that bears a lipstick's traces . . .'. And the 1959 song, *Lipstick on your collar*, written by Edna Lewis and George Goehring, even combined it with a *frisson* of deceit. Today, American women spend almost $700 million a year on lipstick, and world-wide the total spent is probably more than double this amount.

Earlier generations of women also used to enhanced the colour of their lips. Probably the first to do so were the ancient Egyptians, and they used the plant dye henna. An alternative pigment was a reddish-purple dye called fucus, also made from plant material. Some women appear to have even applied cinnabar, the intensely red pigment that painters used (including the cave painters of 20,000 years ago), but this was not to be recommended because it was poisonous mercury oxide.

The first lipsticks to be sold in push-up cases were made in 1915 and were manufactured by Maurice Levy in the United States. They were coloured with carmine, a natural dye extracted from cochineal, which is produced by a small red insect called *Dactylopus coccus* that breeds on a species of cactus native to Mexico. (The female of the species produces the dye.) The trouble with these early lipsticks was that they were not indelible, in other words they stained everything they came in contact with, leaving tell-tale traces on cups, cheeks, cigarettes—and collars. The answer was to use colouring agents that would *dye* the skin red; these were introduced in 1925 and remained popular for a whole generation of women, to be ousted only when their daughters rejected bright-red lips in the 1960s.

The requirements of a perfect lipstick are technically quite demanding. It should produce the desired colour and effect, be it matt, shiny, glossy, frosted, or pearlescent. Lipstick should cover evenly and not feel greasy, have a neutral taste, be long lasting and, of course, be nearly indelible. The stick itself should be smooth to apply, whether conditions are warm or cool; it should retain its shape and not break; it should not be affected by moisture or air, not harbour germs, and not be made from toxic or health-risking materials: quite a demanding set of criteria. Not all these requirements have been met, though most of them have.

Lips are quite a vulnerable part of the human body. Their skin is covered by a thin corneal layer that contains little fatty tissue, and as a result it easily dries out. Normally the moisture is replaced simply by the action of licking the lips, but even this cannot cope with particularly harsh drying conditions such as are encountered in extremes of climate. Then it is best to coat the lips with an oily grease derived from a plant, an animal, or a chemical company. Lip balms for this purpose contain things like castor oil, which comes from plants, or lanolin, which comes from sheep's wool, or petroleum jelly (for example Vaseline), which is a by-product of oil refining, or silicones, which are manufactured by the chemical industry.

A typical lipstick might have the following composition:[2]

Dye	5%
Titanium dioxide	10%
Oil	40%
Wax	20%
Emollient	25%

The oil and the wax are chosen so that together they provide a softly yielding salve for the lips, while allowing the lipstick to remain firm within its cartridge case. The lipstick should also remain solid at temperatures up to 50 °C. All kinds of oils have been used in lipstick, including **natural oils** (see Glossary) such as olive oil and cocoa butter, and mineral oil (also known as liquid paraffin), which comes from the petrochemical industry. Today, the most likely oil is castor oil, which also has the added advantage of forming a tough, shiny film when it dries after application. Almost half the weight of some brands of lipstick can be purified castor oil. Alternatives to castor oil are produced by the chemical industry and these are colourless, odourless, non-toxic, and non-greasy.

Wax (see Glossary) is needed in lipstick to give it shape, and typical waxes are beeswax, carnauba wax, and candelilla wax. Beeswax is the preferred one, and chemically this is a mixture of cerotic acid and myricin, and melts at 63 °C. Over the centuries it has been widely used in furniture polishes, candles, and even in medicines. Carnauba wax oozes from the leaves of the South American pine (*Copernicia prunifera*) and is also known as Brazil wax. This wax is harder and melts at 87 °C, and it too was formerly used in polishes, candles, and as a waterproofing agent, and is still the basis of some car waxes. Its main ingredient is carnaubyl alcohol. Candelilla wax is produced in Mexico and extracted from the candelilla plant (*Pedilanthus macrocarpus*). It melts at 67 °C and tends to be used only when beeswax and carnauba are too expensive. This was the wax

---

[2] There are also traces of other components: some to impart a pleasant odour, a preservative to kill microbes, vitamin E, sunscreens, and sometimes even a flavouring agent.

once used as sealing wax, and is still used on state documents because it will retain the imprint of an official stamp.

Lanolin, which is extracted from sheep's wool, was once thought to provoke an allergic reaction in some people. This now appears not to be the case and it has again become a key ingredient in some cosmetic products, though rarely mentioned because of the previous media scares surrounding it. Some lip-glosses can have more than 70% lanolin in their composition, while eye shadow can have more than 50%, and lipstick more than 25%. Indeed, a few per cent of lanolin is now added to various cleansing creams, foundation creams, hand-cleaner creams, night creams, shampoos, and sunscreens. Its future is not assured because some still doubt its safety, and there is evidence that it causes contact dermatitis in certain individuals.

Colour in a lipstick—most often a shade of pink or red—is provided either by a pigment or by a dye. The most commonly used dyes are D&C Orange No. 5 and D&C Red No. 22. The D&C refers to the Drugs and Cosmetics list of dyes approved by the United States' Food and Drugs Administration (FDA). The chemical names for these dyes are 4',5'-dibromofluorescein and 2',4',5',7'-tetrabromofluorescein, respectively, the former having two bromine atoms in the molecule (which is what the prefix 'di' refers to), and the latter having four bromine atoms (which is what the prefix 'tetra' refers to). The latter dye is also known by the slightly simpler name of eosin.

These dyes are made from fluorescein, itself a yellow dye, which, when reacted with bromine, adds two atoms of this element and turns orange. Add two more bromines and it becomes eosin, which is red, with a slight bluish tinge. The colour can be intensified by converting it into a *lake*, which is the technical term for a dye that has been absorbed on to the surface of inorganic materials such as alumina. The final colour that the dye takes is determined by its chemical reaction with the protein of the skin. The amino groups of the protein bind to the dye and not only make it almost indelible, but also turn it a deep red.

Titanium dioxide is put into lipstick for the same reason that it is added to paints, because of its superb covering ability but, in the case of lipstick, its whiteness serves a second purpose in that it can dilute the colour of the red dyes to give various shades of pink. Of course, not all lipsticks are red or pink. In the search for ways to attract attention in a world of endless distractions, and perhaps even to shock, some women – and men—have painted their lips every shade of the rainbow, and even black.

The chemists employed by makers of lipsticks have struggled hard to overcome the lack of new dyes. Although all kinds of wonderful colours are available, the problem is in getting them approved by agencies such as the FDA. Rather than spend years of testing to prove a new dye is absolutely safe, a risky and costly procedure that can easily fail, chemists now rely on technology to produce the shades and effects that fashion dictates. Interference pigments can be used to create almost any colour and these rely on the way they reflect light when they are deposited on the surface of a titanium dioxide particle. Light striking such a particle can be reflected, refracted, and scattered in ways that are determined by its surface, rather than by the inherent colour of the dye molecule.

The addition of tiny spherical particles, known as microspheres, can also improve the texture of lipstick, making it less greasy as well as more reflective. These tiny spheres are made of the polymer polymethyl-methacrylate, which encapsulates and slowly releases other ingredients such as vitamin E, folic acid, and fluoropolymers, all of which are said to improve the skin. The last of these also has benefits during manufacturing because it not only gives the lipstick a smooth, elegant feel, but also prevents it from sticking to moulds.

To produce a lipstick, the various ingredients are heated and stirred until they are fully blended, and the hot liquid is poured into metal moulds which are then chilled to release the stick. The stick is flamed for half a second to give it a smooth and glossy finish and to remove any imperfections. Pearlized lipsticks include boron nitride (see below), which imparts a shimmer and lustre to the lips. Some include mica or silica particles to impart an extra shimmer. Matt lipsticks have more wax and pigment, giving them more texture and less shine. Glossy lipsticks contain more oil and less wax, which may make them too soft to form a solid stick, and then the product is sold in small pots as lip-gloss. Long-lasting lipsticks usually include silicone oil which seals the colour.

Lipstick has come a long way in the past ninety years and, thanks to chemistry, it is now seen as an indispensable item of female attractiveness, which is sometimes designed to lead on to that other type of chemistry between two people. But while a woman's lips may speak one language, the rest of her face may be telling a less attractive story, and it is then that she might well turn once again to the science of chemistry.

Anti-ageing creams: alpha-hydroxy acids (AHAs)

AHAs are generally called natural fruit acids by advertisers of anti-wrinkle cosmetics, and they really can improve the skin by penetrating the outer layer and stimulating the growth of new skin. Whether they really remove wrinkles is debatable but they may, as the advertisers so guardedly put it, 'reduce the appearance of wrinkles'.

If your career has been mainly out-of-doors, then your skin will be extremely wrinkled when you are old, as is the case among farmers, construction workers, gardeners, and so on. The same will be true of those who elected to spend a lot of time in the sun, enjoying leisure pursuits and the like. That brave band of sun-worshippers of the 1960s and '70s may well now rue their days of carefree exposure. For them it's not a case of no more wrinkles, but of even more wrinkles.

As we get older, the bloom of youth fades. Our skin gets thinner, loses its elasticity, becomes drier, and develops wrinkles. What we want is something to rub on it that will restore its smoothness, add tautness, and remove the lines of age. Every year, new products are launched claiming to do just that, and sales of cosmetics and toiletries now exceed $30 billion a year in the United States, and probably more than twice that world-wide, their sales increasing in line with the world's ageing population. The cynic will say that this is mostly money wasted, that nothing can reverse the slow deterioration of the skin. They would be surprised to learn, however, that some anti-ageing creams really do work, albeit to a limited extent.

Alpha-hydroxy acids (see Glossary) are the active ingredients in many products, and they really help in the struggle to keep the face looking young. But when it comes to advertising claims that they will *eliminate* wrinkles, it is less a question of removing lines and more one of reading between them. Yet we should not allow the hype that surrounds these products to blind us to the help that they can bring.

Nowadays, cosmetic chemists are somewhat hampered in their researches by the health, safety, and environmental laws which make testing of all new chemicals prohibitively expensive. As with the dyes for lipsticks, they now look for other ways to make existing materials more effective. Even so, they are soon to have both hands tied behind their backs in Europe by new laws. In 2003, the European Parliament approved an immediate ban on testing cosmetic products on animals, and on the sales of such products that come from outside the European Union. In 2009 a complete ban will be imposed on cosmetics, any of the ingredients of which have ever been tested on animals, from being manufactured within the EU as well.

There are ways, other than using chemicals, to remove the signs of ageing, such as cosmetic surgery, laser treatments, or botox injections. In the first of these the offending skin is cut away, the second burns off wrinkles, especially those around the nose, mouth, and forehead, while the third method paralyses facial muscles and so prevents skin from creasing. While such expensive treatments are popular with celebrities and actors, whose faces really are their fortunes, they are not something that most people feel the need for, although botox treatments are now easily accessible. These involve the injection of the chemical produced by *Botulinum*, the food-poisoning bacterium, and a treatment generally lasts for several months. (Surgery, laser-burning, and botox are not risk free, but it may be that the only testing done on these wrinkle-reducing treatments was carried out on human animals, so that's OK.)

Botox is also used to treat medical disorders such as cervical dystonia (which causes severe neck and shoulder contractions), excessive sweating, and to cure people who were cross-eyed, which was one of the first treatments for which it received official approval in the United States. Indeed, its effect on wrinkles was first noticed when a Canadian eye specialist, Jean Carruthers, used it to relieve eye twitching in a patient and noticed that her patient's wrinkles had also disappeared. Botox also holds promise of a long-term treatment for migraine and for controlling the spasms associated with Parkinson's disease. Botox works by blocking the release of acetylcholine, the neurotransmitter that causes muscles to contract, but there are side effects associated with its use if the drug migrates to surrounding tissue. For example, it can cause swallowing difficulties in those given injections for neck conditions.

An alternative way of looking younger is to lift off the top layer of dead skin cells at the surface; this can be done by chemical means. While alkaline solutions can be very effective, they are too dangerous to be sold to, or used by the public, so it is to acids that most people turn to achieve these results. As the top layer of skin comes away, it removes superficial blemishes and fine lines, at least for the time being. When done under medical supervision, the result is the formation of a thicker crust of dead skin which takes perhaps a day to form and this is then washed away with soap and water to leave the bloom of new skin. This leaves the face looking somewhat red, but as the days pass this fades, leaving a fresh complexion that will last several weeks.

Less drastic treatment with these acids is available for home use, and some find it reassuring to know that these are the acids that plants produce or which are to be found in milk.

> They say that milk improves the skin,
> But drink it dear, don't rub it in!

So went the old homily, but whoever coined that rhyming couplet was being too clever by half, and poking fun at a traditional method of making the skin look younger by bathing it with milk. Milkmaids of old were renowned for their fresh complexions, and the noted beauty, the queen of Egypt, Cleopatra (69–30 BC), was reputed to have bathed in asses' milk. Nor was she behaving foolishly, and neither were those who used lemon juice on their skin. Milk and lemons, and lots of other natural materials, contain alpha-hydroxy acids (AHAs), and these can have a beneficial effect by helping to remove the top layer of skin.

The most abundant natural sources of the AHAs are sugar cane, which contains glycolic acid, and milk, which has lactic acid, while grapes provide tartaric acid, lemons have citric acid, apples generate malic acid, and bitter almonds yield mandelic acid.

In 1984, dermatologist Eugene Van Scott of Pennsylvania decided to use concentrated AHAs, and chose the simplest of these, glycolic acid, to treat the skin of twenty-seven women clients twice daily for three months. The results were remarkable. Two-thirds of the women reported a noticeable lessening of wrinkles. Further treatments by Van Scott and associates were reported in *The Journal of the American Academy of Dermatology* in 1986, and these showed that after six months of such treatment the skin had become thicker and its elasticity had improved.

In the early 1990s, other fruit acids began to be used in anti-wrinkle treatments, as well as being added to skin creams. Some were indeed from natural sources, such as the cocktail of AHAs produced by the fermentation of fruits. Chardonnay grapes, for example, give a mixture of lactic, malic, and tartaric alpha-hydroxy acids, along with pyruvic acid and acetic acid. Lemons, grapefruits, tomatoes, and bilberries give other mixtures of these acids, while for a more exotic blend some producers have turned to pineapple or passionflower fruits, or used fruits and berries found in the Swiss Alps to bring a hint of clean mountain air—at least in their advertising. One company, Optima Chemicals, sells a product known as SeaAcid, which contains AHAs derived from the fermentation of marine algae, such as seaweed. SeaAcid is mainly lactic acid, with some malic acid and pyruvic acid, but it also contains carbohydrates that improve the feel of the product when on the skin.

While fruit acids can be extracted from cultivated plants, the best source may be chemical plants. Not only are supplies more abundant, and cheaper, but they are purer and less likely to contain other ingredients that might cause an allergic reaction in some individuals. Whatever their source, and whatever their unique selling point, the active ingredients are the alpha-hydroxy acids.

When it was first discovered that AHAs were capable of doing this, there was a rush to market them. And while the results they produced *looked* good, their use was not without risk because the creams contained very high levels of these acids, and in the hands of unskilled users, these concentrations could damage the newly exposed skin as well. In 1989 the US Food and Drug Administration began to take an interest in these products, following complaints from users who said they caused redness, swelling around the eyes, blistering, rashes, itching, and even bleeding.

In 1997 the US National Toxicology Program of the National Institute of Environmental Science also investigated them, and stricter controls were introduced limiting their use to 10% concentrations, and the acidity of the preparations had to be buffered to **pH** 3.5 (see Glossary). Today it is accepted that creams sold to the public should have no more than 8% AHA, but even so they can be effective. An industry-sponsored study found that a group of people who used a 4% solution of glycolic acid twice daily for twelve weeks experienced almost no redness. Nevertheless, in the United States a serious warning was issued in the summer of 2002 as a result of research commissioned by the US Food and Drugs Administration, which showed that skin treated with AHAs was more sensitive to sunburn.

In Britain, some users claimed that AHAs had damaged their skin. A Liverpool-based lawyer coordinated their appeals for compensation and brought their plight to the attention of the media in 1995. Two companies in particular, Clinique and Elizabeth Arden, were targeted, and indeed some brands were withdrawn from the market, following complaints that they particularly irritated the eyes, causing blurred vision in some cases. No major court actions resulted from these media alarms although some out-of-court settlements were made.

Glycolic acid

Chemically this is the simplest AHA and, while it is abundant in sugar-cane juice, it is also present in artichokes, onions, sugar beet, wheat, apple juice, soy sauce, and Chardonnay grapes. Because it is so common in so many foods, it is deemed to be inherently safe. Were it not so, then the toxicity tests carried out on it might have given rise to concern. For example, when glycolic acid was fed to rats it resulted in stunted growth and kidney malfunction. The rats were, in fact, albino rats and the glycolic acid made up 2% of their diet, so perhaps that was the reason it affected them adversely. Nevertheless, such findings would automatically rule out a new material designed for human use, however beneficial it was at removing wrinkles.

Glycolic acid is sold by its manufacturers as a 70% solution. The acid is used in all kinds of ways: it is especially good for cleaning copper; it acts as a scale preventer in water treatment; it is employed in dyeing; and it is used in the manufacture of various compounds such as adhesives. A derivative of glycolic acid, benzyl glycolate, acts as an insect repellent against mosquitoes.

Industrial-strength solutions of glycolic acid have been used medically to repair the ravages of acne and to treat eczema, but can only be applied by skilled dermatologists and doctors. Those employed in beauty salons are more likely to use a 10–20% solution, while that which is sold to the public is likely to be less than 10%.

When pure, glycolic acid exists as colourless crystals but, dissolved in water, it gives an acidic solution which is adjusted to give a pH of around 3. When glycolic acid crystals are heated, they melt at 80 °C and the glycolic acid molecules begin to react with one another to form the polymer polyglycolide. This, too, has its uses, especially in medical devices such as sutures, because it slowly disintegrates back to the acid and the acid is easily excreted from the body. This reversion to the acid can also be put to use when it is essential to maintain an acidic environment over a long period of time.

Is there any real *science* behind the use of glycolic acid for wrinkle reduction? Perhaps. This acid is particularly able to penetrate through cell walls because it is the smallest molecule of the AHAs. Inside the cell, it then stimulates protein metabolism, with the formation of collagen and other compounds, and this plumps up the cell and makes it appear less aged. What is much more evident is that the acid encourages loss of the outer layer of dead skin, thereby exposing, new, younger-looking skin. Higher concentrations, as used in specialist treatments, may even lead to the synthesis of new collagen.

The manufacturers of cosmetic creams containing glycolic acid are recommended to include no more than 4%, although to be effective it has to be double this. If you are tempted to use cosmetics with higher levels than this, then test them on a small area of skin first, and if they cause redness, itching, or stinging then they are not for you.

Lactic acid

Lactic acid is a milder acid than glycolic acid. It can be produced from any fermentable carbohydrate such as whey, dextrose, starch, or molasses, and it can be manufactured from the industrial chemical acrylonitrile, which is used to make all kinds of plastics and fibres. Lactic acid was one of the first organic acids to be discovered and was reported, along with citric acid, oxalic acid, and tartaric acid, by the Swedish chemist, Carl Scheele (1742–86) in the 1780s. Lactic acid is present in milk, gets its name from *lac*, the Latin word for milk, and was once known as milk acid. Lactic acid is also present in bread, cheese, meats, beers, and wines.

Lactic acid is present in the human body in large amounts as a perfectly natural by-product of the metabolism of carbohydrates. (In fact, we emit a little lactic acid on our breath and this is what mosquitoes use to locate us.) Lactic acid is a component of human skin, being the major water-soluble acid in the epidermis, and essential to keeping it in good condition. Consequently this acid is added to skin preparations in the belief that it will compensate if it is naturally lacking. Lactic acid helps sun-damaged skin to heal, and reduces fine lines, wrinkles, and liver spots (the popular name for the brown blemishes of excess melanin pigment that come with age).[3] Lactic acid is sometimes added to shaving creams, and is used to bleach freckles.

Lactic acid works by penetrating the skin and weakening the **hydrogen bonds** (see Glossary) that hold cells together, thereby making it easy to remove the outer layer of dead skin. It also increases and maintains the level of hydration in the upper layer of skin by improving its water-holding capacity and stretchability, so counteracting the tendency of the skin to become dry, flaky, and cracked. Lactic acid can be applied to unblock pores, soften the skin, and in general tone it up.

Lactic acid is buffered to stabilize its pH, so that it is neither so strong an acid so that it acts as an irritant, nor too weak to be effective. Most cosmetic AHA formulations have a pH range of 3–4, which is thought to confer most benefit. Buffering keeps the pH stable and is done by mixing it with the salt, sodium lactate. (See Glossary for more on **buffers**.)

Salt lactates are also used in toiletries: potassium lactate, for example, increases viscosity and makes what are referred to as 'thick' liquids. Silver lactate has been used in anti-dandruff shampoos, and zirconium lactate makes a good antiperspirant.

There are other kinds of lactates that find their way into cosmetics. These are lactic acid esters, such as butyl lactate, lauryl lactate, and myristyl lactate, which leaves the skin feeling soft and smooth. (See Glossary for more on **esters**.) One ester, ethyl lactate, is destined for even wider use—see box on p. 14.

Lactic acid has a future beyond these cosmetic preparations. It is destined to play a bigger role during this century as the chemical industry converts to using renewable resources. It will enter our lives as the polymer polylactic acid (PLA), which is now being produced from agricultural crops in the United States. A plant at Blair in Nebraska has been built jointly by the US company Cargill and the chemical giant Dow Chemicals to make lactic acid from crops such as corn or sugar beet. The acid is then turned into its polymeric form, which is suitable for making

---

[3] Their medical name is senile lentigo.

Ethyl lactate, a renewable solvent for industry

Industrial cleaning requires solvents, and these should be efficient, safe, and effective. A solvent that evaporates into the atmosphere, and whose vapour does not rapidly decompose, will act as a greenhouse gas. Solvents should come from renewable resources; in other words, those that are currently derived from fossil fuels will one day have to come from agriculture. Ethyl lactate meets these requirements.

In the United States, Holland, and Spain, there are already chemical plants making ethyl lactate from agricultural sources, and there is already a world-wide demand for this solvent of almost 20,000 tonnes a year, even though it costs around €4 per litre, which is four times the cost of other cleaning solvents. In the United States, ethyl lactate is made from cornstarch, while in Holland and Spain it is made from sugar beet. The lactic acid is made by fermenting the starch or sugar using bacteria or fungi.

Ethyl lactate (also known as Vertec) is a colourless liquid with a characteristic fruity odour. It boils at 145 °C, which means that it is not very volatile so less is lost to the environment, and that which does evaporate is quickly degraded. Vertec has the ability to dissolve a wide range of materials and has already replaced solvents such as xylene, acetone, and dimethylformamide.

packaging and textiles, and sold under the trade name of NatureWorks. The solid form of PLA resembles polystyrene, while its transparent film is like cellophane, and its fibre is similar to polyester and suitable for making T-shirts and carpets.

So do AHAs really work? The answer would appear to be yes. A double blind test by Matthew Stiller was reported in the journal *Archives of Dermatology* in 1996. It looked at the effect of 8% AHA solutions on seventy-four women of ages ranging from forty to seventy who all had skins that had been badly damaged by overexposure to the sun earlier in their lives. Stiller concluded that the daily application of AHAs over a period of six months produced clear benefits, although they had limited anti-ageing effects.

Lynn Drake of the University of Oklahoma reported on AHAs at the 55th Annual Meeting of the American Academy of Dermatology in 1997.

She said that she had carried out double-blind tests on creams with both 8% glycolic and lactic acids, and found them to be 'slightly, but significantly' better at improving the appearance of skin, especially when this had suffered damage due to overexposure to the sun.

Finally, there are the beta-hydroxy acids (BHAs)—see under AHAs in the Glossary—which, like their alpha cousins, also act as exfoliants, but it is said that BHAs are better at reducing fine lines and wrinkles without the irritation that sometimes accompanies the use of AHAs. There is no convincing evidence as yet that BHAs are better than AHAs.

Silky smooth skin: boron nitride

It may have been developed for use in the aerospace industry, and is now used industrially to make crucibles for holding molten metal, but it can make ordinary cosmetics feel like silk to the skin.

One of the wittiest product names is surely *Très BN*, which is not only a play on the name of boron nitride (chemical formula BN), but also hints at the chic associated with the French reputation for style and flair, and that's what the use of this rather curious chemical is all about. It is manufactured by Saint-Gobain Advanced Ceramics, formerly known as the Carborundum Corporation, which gives a hint as to the purposes that it was originally designed for. Somewhat surprisingly, boron nitride was initially developed to meet the needs of the aerospace industry and the extreme conditions that high-powered flight entails.

Boron nitride was first made in 1842 by a chemist called Balmain, who heated together a mixture of boron oxide and mercury cyanide. Better methods of making it were soon discovered, such as heating sodium borate (borax) with ammonium chloride, and when BN became a commercially saleable product, it was prepared by heating borax and melamine in an electric furnace. Boron nitride consists of a three-dimensional array of alternate boron and nitrogen atoms chemically bonded together, and it can exist in two forms that mimic those of pure carbon, that is, graphite and diamond. A chemist finds this is not too surprising, because the combination of boron (5 electrons) with nitrogen (7) has the same electronic arrangement as carbon (6) with carbon (6).

In graphite each carbon is linked to three other carbons in the same plane, forming honeycomb-like platelets, and these platelets stack up one above the other. They can glide easily over one another and this confers a lubricity on graphite that is exploited in certain industrial and engineering greases. The same property is to be found in the graphite-like form of BN, which is known as hexagonal-BN. This, too, is a good lubricant, and it is used in high-temperature situations, such as hot pressing and machining. It also finds use in insulators and crucibles, and even for making the composites used in microwave oven windows.

On the other hand, diamond is far from being soft and flaky, and is the hardest naturally occurring substance known. It has a structure in which every carbon is linked to four other carbons in a pyramidal three-dimensional arrangement, which gives it strength and rigidity. It is this arrangement, too, that is present in the diamond-like form of BN, which is known as cubic-BN, and which comes a close second to diamond in terms of hardness. Like diamond, it is also used as an abrasive, especially for iron and steel cutting tools. The cubic form of BN was first obtained in 1957, and was made from the hexagonal-BN by heating it at 1800 °C under a pressure of 85,000 atmospheres.

When hexagonal-BN is pressed into shape it can easily be machined into complex shapes, and it is not wetted by most molten metals, making it ideal for crucibles. It is also a good electrical insulator, and in that respect it is unlike graphite, which conducts electricity. Though chemically more stable that graphite, hexagonal-BN is susceptible to oxidation, and its use at high temperatures is somewhat restricted.

At this point you might be forgiven for wondering what BN could offer by way of a cosmetic ingredient, and yet it is present in many of them. Hexagonal-BN is white, and when it is ground to a powder it imparts a softness, silkiness, and smoothness, along with a pearly sheen. It is added to foundations, lipsticks, and nail polishes. The traditional way of producing a pearlescent effect was with bismuth oxychloride, $BiOCl$, but hexagonal-BN is better, especially if crystals of this are allowed to grow, enabling them to scatter light to give a glittering or pearlescent finish. The amount of BN in a cosmetic is generally between 1 and 10%, but it can be much higher. It is completely non-toxic and not hazardous in any way. It is used in foundations to cover wrinkles because it reflects light, and it is lack of reflectance, compared to that of the surrounding skin, which makes wrinkles so noticeable.

Hexagonal-BN has a lower coefficient of friction than talcum powder which, for centuries, was the basic material of most cosmetics. When it was reported in the 1990s, however, that talcum powder might be linked to cancer of the ovaries, this material fell out of favour with the manufacturers of cosmetics and toiletries, and they turned instead to hexagonal-BN as a replacement.

Sunblocks, sunscreens, and instant tans[4]

When you want to protect your skin against the sun's ultraviolet rays, which can cause cancer, there are lotions that you can apply with varying degrees of protection. There are also products that enable you to achieve an instant tan, and some that can even make your skin lighter.

The current ideal of the body beautiful is one with a light to moderate tan. It is still fashionable to regard such a tan as healthy, and it certainly looks attractive. The majority of people get their tan when they go on holiday, deliberately choosing to visit sunny locations where the sun can be guaranteed to shine almost every day, and they flock to the coastal regions of the Mediterranean, the Caribbean, Australia, south-east Asia, California, and Florida. In so doing, they put not only their health at risk from skin cancer but also do long-term damage to their skin, which eventually takes on the texture of leather and is politely referred to as 'weather-beaten'. The culprits are ultraviolet rays and mainly those referred to as UV-B. These rays can penetrate deeply and can trigger cancer by damaging DNA and the immune system.

---

[4] A sunblock is an opaque substance that physically blocks the sun's rays from reaching the skin. Zinc oxide and titanium dioxide are sunblocks. A sunscreen is transparent and absorbs the sun's rays by virtue of its chemical nature.

Visible light is made up of six readily distinguishable colours: red, orange, yellow, green, blue, and violet. (It is of course possible to divide the spectrum in much more sophisticated ways, but these colours seem to be the main categories and the ones that are part of everyday language.) Red light spans the wavelengths 740–620 nanometres (nm), orange 620–585 nm, yellow 585–575 nm, green 575–500 nm, blue 500–445 nm, and violet 445–400 nm. Most human eyes do not register wavelengths shorter than 400 nm, but if they could then we would see two more colours: ultraviolet-A, which has wavelengths 400–320 nm, and ultraviolet-B, 320–280 nm. There is a third ultraviolet colour, ultraviolet-C, of wavelengths 280–100 nm, but most does not reach the Earth's surface because almost all of it is absorbed by the ozone in the upper atmosphere.

Although we cannot *see* UV-A and UV-B, our eyes can be damaged by them and our skin reacts to them: UV-A causes it slowly to go brown; UV-B produces a quicker response and it goes a fiery red. UV-A rays are sometimes referred to as the ageing rays because that's the effect they ultimately have, while UV-B rays are called the burning rays because they cause blood vessels near the skin's surface to dilate and carry more blood, thereby making the skin feel hot and look red.

Human skin cells are colourless and they have evolved a mechanism to protect themselves against UV light, which they do by producing special molecules. These are to be found in the outer layer of the skin, the stratum corneum, and they have the ability to absorb UV rays. Typical molecules are the **amino acids** (see Glossary) such as tryptophan and tyrosine, plus urocanic acid, which is produced from the amino acid histidine. Urocanic acid suppresses the immune system and as such is not something we want much of, despite the protection it provides against UV. Before this was realized, urocanic acid was added to cosmetics as a moisturizing agent, and in the 1960s it was regarded as a 'natural' sunscreen, which indeed it is. This use is now banned.

The most effective UV-blocker is a chemical pigment called melanin, which is produced in the upper layers of the skin in cells called melanocytes. (Melanins are also the pigments for the dark colour of hair, feathers, fur, and fungi.) These cells are triggered into producing this black pigment when they are exposed to UV-A rays, and they make it from the amino acid tyrosine, which combines with other molecules to form the melanin polymer. The more UV-A is absorbed, the more melanin is formed, and the deeper the colour of the skin becomes. Other creatures also rely on melanin for protection, including mammals, insects, various plants, fungi, and even micro-organisms. People of African origin are born with naturally high levels of melanin and maintain these levels through life. In fair-skinned individuals, the skin takes time to produce melanin, and so it cannot protect against sunburn following sudden exposure.

People are classified into four skin types, type I being the most sensitive to burning and type IV being the least sensitive. People of Celtic descent from northern Europe are types I and II, and they suffer sunburn easily and tan only poorly because they cannot produce enough melanin fully to protect themselves, whereas people of Mediterranean origin are types III and IV, and they tan easily and burn rarely. Fair-skinned redheads are invariably at risk from sunburn, even though their skin also produces a form of melanin, because their melanin affords little protection. A new drug, Melanotan, has been developed in Australia and may one day be on the market. It can be taken orally and it stimulates the formation of melanin in the skin, thereby allowing people to get a natural tan and natural skin protection. Melanotan is a thousand times more potent at triggering the formation of melanin than the body's own enzyme, and it lasts longer.

Some individuals have so few melanocyte cells that their skin becomes inflamed and red when exposed to levels of sunlight that would not affect other people, and they need to use artificial sunscreens on a regular basis if they want to spend time outdoors. Such people are said to be photosensitive, and this condition can develop as a side effect of taking certain drugs.

Skin cancer is the downside of prolonged sunbathing, but not all exposure to the sun is bad, and time spent outdoors is generally beneficial. The action of UV rays on the skin makes a valuable contribution to the body's need for vitamin D, lack of which in children leads to a weakened skeleton, and thence to rickets, because this vitamin is needed by the body to utilize the calcium from which bone is made. In adults, vitamin D helps maintain the skeleton, prevents clinical depression, and also protects against heart disease. A few minutes of sunshine every day is good for us, and according to researcher Michael Holick of Boston

University, Massachusetts it can even ward off other forms of cancer such as colon, breast, and prostate cancers by boosting the amount of vitamin D in the body. Although we can get vitamin D in our diet, that produced by sunlight falling on the skin appears to stay in the body longer, and there may be other molecules produced along with it that are also beneficial.

Every year in the United States, more than half a million people develop some form of skin cancer, and the American Cancer Society claims that one person in six will experience this disease at some time in their lives. The most common type is basal cell cancer, which forms as raised, hard, red spots on the most exposed parts of the body, that is, on the face, hands, and neck. The next most common is squamous cell cancer, which is a hard-surfaced lump that generally appears on the lips, ears, and hands. Neither a basal cell nor a squamous cell cancer is life threatening because they do not spread to other organs and they are easily removed.

The same cannot be said of the third form of skin cancer, melanoma, which does spread. It often starts as a mole, which then grows in an unusual way and turns a blue-black colour. Thankfully, melanoma is very rare. Nevertheless, it kills about 6000 Americans every year, and about the same number of Europeans, which means that in a moderate-sized town of 100,000 inhabitants there would be one or two cases of melanoma per year compared to around sixty-five cases of the other forms of skin cancer.

But is UV-B really to blame? Perhaps not. It now appears that most of those who get melanoma are predisposed to develop it because of a genetic mutation, which explains why it often starts on parts of the body that rarely see the light of day. The genetic defect was discovered by Richard Wooster and fellow researchers at the Wellcome Trust Sanger Institute in Cambridge, England, in 2002. A few years earlier, Marianne Berwick of the Memorial Sloan-Kettering Cancer Center in New York had suggested that there may not be a link between exposure to sunlight and melanoma. She analysed the many epidemiological studies on the supposed relationship between sunblocks and melanoma, and concluded there was little evidence that they protected against it. Nevertheless, sunscreens do prevent the other forms of skin cancer, and they work in different ways, so we should not be surprised that the total sun-care market is now in excess of $1 billion per year.

There are three ways we can protect our skin against UV rays: reflect the rays away from the body; absorb the rays but deactivate them; or prevent and repair the damage they cause. There are chemicals available to do all three. In the United States and Europe, people spend around $200 million a year on sunscreen and sunblock creams, and most people are now aware that applying sunscreen oil or cream only at the start of the day may not be enough, because it may be sweated off if the wearer engages in vigorous exercise, or it may be washed off while swimming. We may feel refreshingly cool in a pool but this does not mean we are safe from UV rays, because they pass through water unabsorbed. Silicone oils, which are by nature water-repellent and are added to some creams, provide a longer-lasting degree of protection for those who want to exercise in the sun or go swimming.

A good sunscreen should include microfine titanium dioxide or zinc oxide to protect against UV-A, and an organic compound to protect against UV-B. It should be easy to apply to the skin, giving a continuous film, and yet be invisible. It should not be sticky, or easily washed off when swimming, although it should be easily removed using soap or bath gel. It should not harbour germs, and it should smell nice. And finally, of course, it should have a reasonably high sun-protection factor (SPF) of 15, with this remaining high throughout the day.

SPF is an indication of the extent to which a sunscreen will block out UV rays, and is given as a number on a scale ranging from 2 to 30, and beyond. It is calculated by comparing the time that it takes to register sunburn on protected skin compared to the time it takes on unprotected skin. For example, if sunburn can be detected after two hours unprotected but only after twelve hours while wearing a sunscreen, then that sunscreen has an SPF of 12 divided by 2, giving a value of 6. Thus SPF is a measure of the degree to which a sunscreen *reduces* the UV rays reaching the surface of the skin. Factor 2 cuts UV by half, in other words to 50%, a factor 4 allows only a quarter of the rays to penetrate to the skin, so absorbs 75% of the UV, while factor 10 cuts out 90%, and factor 25 cuts out 96%, permitting only 4% to penetrate. In theory, it is possible to have factor 50, which would block 98% of the UV, but while such creams have been produced, they offer little more in the way of protection than factor 30.

Originally, factor 10 was considered high protection, but some wor-

ried consumers now use factor 30 with the result that a two-week holiday, with this sunscreen applied regularly throughout the day, should
give them less than a day's full exposure to UV rays. SPF numbers can be
misleading, and they may suggest to some people that a cream with an
SPF of 30 will give you *twice* the protection of one with an SPF of 15. It
doesn't; it offers you only 4% extra protection. It screens out 97% of the
sun's UV rays, compared to 93% for SPF 15.

A typical sunscreen lotion will consist of 5% of microfine titanium
dioxide, 5% of UV-absorbing chemicals, 10% of various oils, and 5% of
an emulsifying agent, with the rest being distilled water. The emulsifying agent prevents the oil and water separating into two layers. The
lotion will also need a preservative to keep it free from micro-organisms.

The best way to reflect UV rays is to put a layer of zinc oxide (ZnO) or
titanium dioxide (TiO_2) on the skin, and these white pigments are sometimes applied as a paste ('zincy') to the most sensitive areas of the face
such as the nose, cheekbones, and lips. Those who perforce must spend
long hours in the sun, such as farm workers, construction workers, and
sports people, may choose to wear such protective ointments. Less obtru-
sive forms of these ointments are also available, thanks to the skill of
chemists, and these pigments can be suspended in a sunscreen and yet
appear invisible when applied.

The trick with titanium dioxide is to use microfine or **nano-sized** particles (see Glossary) that are invisible to the naked eye, and the optimum
size is around 50 nanometres. Such protection can also be combined
with other UV filters, and indeed they have a synergistic effect, each
boosting the other's power so that less is needed of both. They can work
against one another, however, the organic sunscreen causing the microfine particles to cluster together, thereby reducing their protective effect
as well as making the sunscreen noticeable because it takes on a whit-
ish appearance. Adding silicone oils to sunscreen avoids this by keeping
the titanium and zinc oxides in their microfine form. Another way of
counteracting the whitening effect of titanium dioxide is to coat mica
particles with it, and the result is a more natural skin tone.

The second way of protecting skin is to coat it with a transparent film
that will, like glass, filter out the dangerous UV rays. There are various
kinds of UV absorbers and they defuse the rays in different ways. For
example, ortho-hydroxy-benzophenone absorbs UV radiation with a

wavelength of around 330 nanometres and, as it does so, one of its hydrogen atoms moves from the oxygen atom to which it is chemically bonded to an adjacent oxygen atom. This higher-energy molecule then reverts to its original configuration, and as it does it releases its extra energy as gentle heat.

Other molecules can perform a similar trick, moving an electron to an outer orbit, from which it returns, releasing the absorbed energy by making the molecule vibrate; to perform this trick the molecule needs to contain an arrangement of atoms known as a chromophore.

More than 6000 tonnes of UV-protection agents are produced annually by the chemical industry, and these include chemicals based on aminobenzoic acid, camphor, and cinnamic acid, which are particularly effective against UV-B, while those based on benzophenone and dibenzoylmethane filter out UV-A.

The above types of chemicals are referred to as *organic* sunscreens, so-called because they have carbon-based molecules. As such, they are generally soluble in oils, albeit sometimes not soluble enough for commercial preparations, in which case they need to be 'solubilized' by having hydrocarbon groups attached to them. Even if this is undertaken, however, some organic sunscreens are unsuitable because they discolour on exposure to sunlight, while others diffuse into the plastic of containers.

The third method of protection is to repair the damage that UV rays cause by using a radical scavenger. UV rays have the power to break chemical bonds and in so doing they produce molecular fragments called **free radicals** (see Glossary). A free radical will react violently with almost any other molecule it encounters, and that includes those that compose DNA. The molecule that produces the most deadly free radicals is oxygen (O_2), and the combination of this gas and UV light forms so-called singlet oxygen, which is a highly activated species that is most dangerous.

Free-radical damage to DNA can easily be repaired by the body, but UV light also affects the immune system, whose job it is to detect the damage and put it right. The chance of a mutant strain of DNA escaping detection is thereby increased. The cells of the body produce free-radical scavengers for their defence. Special enzymes mop up free radicals, and there are antioxidants, such as vitamins C and E, which can also do the job. Not surprisingly, these are added as ingredients to sunscreen lotions, along with other antioxidants such as the naturally occurring polyphenols, which are chemicals that can be extracted from products such as grape seeds.

There are claims that sun protection can be achieved by taking BioAstin, a dietary supplement which owes its activity to the chemical astaxanthin, a powerful antioxidant which provides the red colour of shrimps and wild salmon; this is said to be 500 times as strong an antioxidant as vitamin E. Tests on individuals were reputed to show that it reduced the sensitivity to sunburn in half of a group of twenty-one people who took it regularly. Large-scale trials are now underway.

The route to today's sunscreens has not been without false starts and controversy. The first sunscreen cream was launched in the United States in 1928, and it contained benzyl salicylate and benzyl cinnamate, neither of which acts as a particularly effective agent. Other skin creams and lotions were aimed at relieving the painful effects of too much sun. It was the growth, in the 1960s, of package holidays to sunny climes, however, that led to the dangers of too much exposure to the sun's rays beginning to be linked to the increasing incidence of skin cancers.

Sunscreens have to meet certain criteria, and there are lists of prescribed and proscribed ingredients that can and cannot be used in them. Nevertheless there are seventeen sunscreen chemicals that are approved for use in the United States and twenty-five in Europe. The restrictions came about because of the possible allergic reactions by some individuals, and some of the previously used sunscreens, such as the once-popular PABA (*para*-aminobenzoic acid), have been withdrawn for this reason. This particular agent was used in the first truly successful sunscreens, but it was eventually discovered that 1% of those who used them found it made their skin more photosensitive. Tests on human cells showed that, while PABA itself did not cause cancer, it increased the formation of cancer precursors. PABA was originally thought to be a vitamin, one of the B vitamins, with bran, kidney, liver, and yoghurt being good sources. Further studies, however, showed that it was not possible to produce dietary deficiency symptoms by excluding it from the diet and its status as a vitamin was withdrawn.

The most common sunscreen chemical is octyl methoxycinnamate (coded OMC, with trade names such as Parsol or Neo-Heliopan), to which almost no one has been found to be allergic although, when it is mixed with some traditional agents for treating skin diseases, such as Peru Balsam, it can react chemically to form a compound that can produce an allergic response. OMC and another commonly used sunscreen, 4-MBC (short for 4-methylbenzylidene camphor), have been criticized as potential endocrine disrupters by environmental toxicologists. This followed research by Margaret Schlumpf of the University of Zurich, Switzerland, in 2001 which showed that hairless rats which had been immersed in a solution of these compounds in olive oil experienced an increase in weight of the uterus. The previous year, Terje Christsen, of the Norwegian Radiation Protection Authority, near Oslo, showed that OMC killed mouse tissue in much smaller doses than previously thought. The risk to humans from these agents is thought to be minimal because they do not penetrate the outer layer of dead cells of human skin, to which they are applied. The chemical company Merck has found a way of encapsulating OMC agents inside a polymer, so that they never actually come in contact with even the outer layer of skin cells.

Sunscreens can be quite expensive. A survey by the Consumers' Association in the United Kingdom in 2001 showed that complete protection for a two-week holiday in the sun could be as high as £63 for the best-known sunscreen (Ambre Solaire), or as cheap as £31 using a supermarket brand.

So what should we do to protect our skin from the ravages of UV rays? The best advice is to avoid going out in the sun between the hours of 11 a.m. and 3 p.m. Outside these hours, use a sunscreen with an SPF factor of 15 if you want to expose your body on the beach or when swimming out of doors. Better still, if you expect to be out in the sun for a long time, say walking, gardening, or playing sport, then wear some form of clothing and headgear. The best protection for the skin is still clothing, and the German chemical company BASF has even come up with a nylon fibre that has titanium dioxide particles built into it. Closely woven clothing made from this can effectively offer complete protection.

Tanning in the dark: dihydroxyactetone

When certain foods are cooked or left to ferment they turn a lovely golden brown because they undergo what is known as the Maillard reaction, in which carbohydrate reacts with protein.[5] The outer layer of our skin is mainly protein, and it too should be able to react with a carbohydrate and so turn brown. A chemical that can cause this to happen is dihydroxy-acetone (DHA), which is a white crystalline powder with a sweet taste. Apply a solution of this chemical to your skin and it will give you the appearance of a tan although, when such products were first sold, they tended to produce an uneven streaking effect that was more embarrass-ing than the pallid skin they were meant to be colouring; nor could it be washed off. The unlucky user had to wait a week or so for the outer layer of the skin to wear away.

DHA as an artificial tanning aid came to light at the Children's Hospital at the University of Cincinnati in the mid-1950s. Here, Eva Wittgenstein was investigating children who were suffering from an inability to store glycogen, the essential carbohydrate that the body uses as a readily available supply of glucose. She believed that this condition could be treated with large doses of DHA. This molecule is produced in plants and animals as an intermediate in carbohydrate metabolism, and is more rapidly assimilated than glucose itself. Unfortunately, some of the children vomited it back, and it was noticed that where it fell on exposed skin this turned brown after a few hours. Wittgenstein was in-trigued and tested her own skin with aqueous solutions of DHA, and indeed she too turned brown.

DHA is prepared industrially by the fermentation of glycerol (gly-cerine) by *Acetobacter suboxydans*. This microbe removes two hydrogen atoms from the molecule, and so converts it to DHA. DHA will combine with free amino groups in the proteins of the stratum corneum, and turn them brown within thirty minutes; the compounds being formed are known as melanoidins. This type of tan is limited to the outer layer of skin.

There are various tanning lotions and creams on the market, and these contain between 2% and 5% DHA. They have to be slightly acidic and are **buffered** (see Glossary) by adding phosphate to give a pH of 5. The com-monly available instant tanning lotions, such as Man-Tan, Oxatone, Q.T. (Quick Tan), and Winter Tan, are of this type. The problem with DHA is that, although it mixes readily with water, it tends to form a gel and this

[5] The Maillard reaction is named after the French chemist, Louis-Camille Maillard, who first described it in 1912.

does not spread easily and evenly on the skin, which is why it can lead to a streaky appearance. This can be overcome by adding other chemicals, either to keep the lotion free-flowing or to form a cream in which the DHA is properly dispersed. In some beauty salons, self-tanning agents are also applied in spray form using an airbrush.

Self-tanning products are not really sunscreens, although they confer some protection with an SPF of around 2. They are basically designed to give people the golden tan that they desire without having to expose themselves to UV rays to achieve it. Of course if you want to fool those on the beach with a ready-made tan, and yet still be properly protected from the sun, then there are preparations which combine DHA and titanium dioxide; these can have SPFs of 15, and even 30.

Skin-lightening creams

Just as it is possible to use chemicals to produce a fake tan, so there are chemicals that will make the skin appear lighter. One of the signs of ageing is the appearance of small patches of melanin known as liver spots. To stop these forming, the metabolism that produces them must be blocked, and that means inhibiting the *tyrosinase* enzymes (also known as *phenol oxidase*) that convert tyrosine to melanin. *Tyrosinase* is widely distributed in Nature, being present in plants, animals, and humans. It is the enzyme that causes a cut apple quickly to go brown. *Tyrosinase* has a copper atom at its active site, and if this can be deactivated in some way then the enzyme loses its oxidative function. This is why, if you cut an apple and then dampen it with lemon juice, it remains white because the vitamin C of the juice deactivates the enzyme. And that is why lemon juice was traditionally used as a skin-whitening agent.

Other conditions that might lead a person to use a skin-lightening lotion are excessive freckles, pregnancy marks, and even too much of a tan. Dark skin is also associated with some medical conditions.

A typical skin-lightening cream is likely to contain known *tyrosinase* inhibitors, such as kojic acid and hydroquinone, together with glycolic acid which removes surface layers of skin so that the active agents can reach the target cells. It will also contain a preservative and various other ingredients such as solvents, vitamins, and fragrances.

In the past, dark skin could easily be lightened by the application of hydroquinone, a chemical commonly used in photography, but this is now discouraged as a skin-lightening agent because some people were adversely affected by it. Hydroquinone is also known by the older name of quinol, and it was marketed under several trade names such as Aida, Black and White Bleaching Cream, Eldopaque, and Tecquinol.

Hydroquinone comes as white crystals which melt at 171 °C, and which are soluble in water and in alcohol. At low concentrations, its solutions would bleach the skin, but in some people it caused dermatitis, and is now illegal in Europe, the United States, and Japan. It can still be used in some countries, but the amount must not exceed 2%, and at such a low concentration it is on the verge of being ineffective. Similar but safer chemicals, such as THPOP (short for 4-(tetrahydropyran-2-yl)oxyphenol), have been patented as skin-lightening agents but are not yet on the market.

There are several natural skin-lightening chemicals: kojic acid, liquorice extract, scutellaria extract, and mulberry; they also inhibit *tyrosinase*. Kojic acid,[6] which is produced by the microbe *Aspergillus oryzae*, is a skin lightener and an antibiotic, and it is also manufactured industrially to be converted to maltol, which imparts a 'freshly baked' odour to bread and cakes. This and the other whitening agents are almost as good as, or even better than, hydroquinone, or so it would appear when their relative abilities to deactivate the enzyme are compared. A 5% solution of hydroquinone will inhibit the activity of *tyrosinase* by 50%, whereas it requires a 10% solution of kojic acid to achieve this reduction, but only a 0.5% solution of mulberry extract.

With all these natural chemicals, treatment has to be continued for several weeks to produce a noticeable effect, and they need the presence of a substance that will aid penetration of the skin if they are to be effective. In addition, they require a buffering agent to keep the solution at around pH 4 and an antioxidant, such as sodium metabisulfite, to prevent the mixture from going off.

In parts of Africa, people with very dark skins sometimes use mercury iodide soap to lighten their skin or to hide dark marks and blemishes; dark skin scars very easily. People using such soaps were advised to leave its lather on their skin all night. The reason for banning it stems from the danger the mercury poses to health, and to the health of those who

---

[6] Chemical name: 5-hydroxy-2-(hydroxmethyl)-4-pyrone; elemental formula: $C_6H_6O_4$.

manufacture these soaps. In the 1970s, tests on workers in such factories showed high levels of mercury in their blood, and, at the University of Lausanne in Switzerland, tests on users of such soap showed that they also had raised mercury levels in their blood and urine, indicating that the metal was being absorbed through the skin.

Liposomes

Skin relies on a substance called collagen to give it its elasticity. Aged and sun-damaged skin loses collagen and becomes slacker and less elastic. Other things also begin to be inadequately produced, and the skin becomes drier and lacking in natural oils, vitamins, and even minerals. Help is at hand in the form of creams and lotions that replace some of the missing components; these work by using liposomes.

A typical moisturizer contains the following: water (which is generally listed under its Latin name of *aqua*), humectants, occlusives, liposomes, antioxidants, and preservatives. Water is present because that is what your skin always needs, and humectants help it to retain this water. Common humectants are the traditional **glycerol** (see Glossary), and a more modern one, sorbitol. Occlusives act as a barrier to prevent water loss, and good ones are lanolin, which comes from raw wool, and petroleum jelly, which comes from oil refineries and is better known by its trade name of Vaseline.

Liposomes are the preferred way of delivering molecules to the layers below the surface of the skin, and especially the layer beneath the stratum corneum. Liposomes are tiny spheres that encapsulate an active ingredient and carry it to the site where it is needed, especially to parts of the epidermis to which otherwise the ingredient could not penetrate, and in this way it has been possible to deliver anaesthetics, vitamins, hormones, and steroids to the lower layers of the skin. Liposomes have an outer layer of phospholipid molecules of the sort that form the membranes of the cells of our body. The phospholipid consists of a glycerol molecule to which are attached two long-chain fatty acids and a phosphate group. A choline group, which has a positive nitrogen as part of its structure, is also attached to the phosphate, which carries a negatively charged oxygen atom. These charged atoms give the phospholipid the ability to interact to form a film of material in which they are on the inside and the fatty acids are on the outside, thereby making a waterproof membrane about four nanometres thick. The best-known phospholipid is lecithin, which is produced on a large scale by plants and animals and is extracted from egg yolk and soya beans. It is widely used in the food industries.

Liposomes are **nano-sized** droplets (see Glossary), typically 100 nanometres in diameter (which is a ten-millionth of a metre and so invisible to the naked eye), and each can carry billions of the active agent molecules. They penetrate through the surface layers of the skin, and then slowly release the active agents through the liposome membrane, or rapidly if the membrane is ruptured, and they are ideal for transporting difficult-to-dissolve materials to where they are needed.

The ability of liposomes to trap molecules and ions was first reported by Alec Bangham at the Institute of Animal Physiology in Babraham, near Cambridge, England, in the early 1960s. Indeed, liposomes became scientific models for cell membranes. They are remarkably stable entities, and ideal for holding other chemicals. Phospholipids can arrange themselves into spheres ranging from 20 nanometres in diameter up to 100 microns, that is, 5000 times larger. The smaller ones are sometimes referred to as nanosomes.

Liposomes are made by evaporating a solution of the lipid in a solvent, such as chloroform and methanol, to form a film. When the film is then put in water, and vigorously stirred, it swells and breaks up, and the broken pieces curl up into spheres which trap the desired molecules inside them. The spheres can be tailored to the required size by blasting them with ultrasound, or by forcing them under high pressure through tiny pores, whereupon the larger liposomes reform themselves into smaller spheres.

The cosmetic industry makes its liposomes from uncharged lipids rather than from the natural phospholipids. These so-called 'non-polar' lipids are dissolved in ethanol and homogenized before being filtered through pores of decreasing size until they are about 200 nanometres in diameter, this being the optimum size for penetrating the skin.

Even if liposomes were to deliver only water to the layers below the stratum corneum, that in itself would produce notable benefits by moisturizing the skin, thereby making it less dry and more elastic. If the liposomes carry antioxidant vitamins, such as A, C, and E, then so much the better as these counteract the free radicals that are thought to be particularly damaging. Of these, vitamin A (also called retinol) has received most attention as an anti-wrinkle agent, and it can indeed make the skin seem younger by reducing large wrinkles, brown spots, and surface roughness.

In the past, molecules related to vitamin A, such as tretinoin, have been used with some success to treat acne, and are now more widely used to delay skin ageing. (Tretinoin is also known as vitamin A acid, or retinoic acid.) The commercial names for tretinoin are Retinova in the United Kingdom, and Retin-A in the United States where it was approved as a wrinkle treatment by the Food and Drugs Administration (FDA) in the mid-1990s. Tretinoin was introduced in the 1970s as the standard treatment for acne, but those who used it noted that it also caused wrinkles to disappear. Over-the-counter versions are available, and these will generally causes age spots and wrinkles to start to fade after two months and be significantly reduced after six months. Tretinoin works by stimulating production of collagen and preventing it from being destroyed, and its benefits can last several years.

Naturally, liposomes attracted the attention of cosmetic chemists in the large companies, and L'Oreal and Christian Dior introduced liposome anti-ageing skin formulas in 1987. To begin with these were exclusive products and priced accordingly, but they were soon joined by competitors such as Nivea who were able to offer the benefits of liposomes to supermarket shoppers.

Price is no guide to performance . . .

In 2001, the UK consumer magazine *Which?* carried out a survey of moisturizers to see whether these improved the texture of skin. Thirty-two women between the ages of twenty-five and fifty-four took part, and every woman tested twelve products each for a five-day period per product. The conclusion of this somewhat limited test was that they generally did improve the skin although there was little relationship between their effectiveness and their cost. A typical jar of moisturizer contains 50 ml, and prices ranged from around £3 to £26. The ones voted the best were Boots No7 (and at £3.15, the cheapest), a medium-priced product, Synergie Wrinkle Lift (£7.99), and an expensive one, Oil of Olay, which cost £18.50. Needless to say, the ingredients in such products rarely justify a high price, although heavy advertising and expensive packaging perhaps can.

. . . nor are popular misconceptions

There are those who decry all things chemical as potentially dangerous, while extolling the benefits of all things they describe as natural, meaning they come from a biological source such as a plant . Chemically there is nothing to choose between them, and your skin is indifferent to the source of the creams and lotions you use to rejuvenate or protect it. In some ways, the chemically sourced material is likely to be better because it is quality controlled and free from traces of extraneous material to which you might be allergic. For example, natural rubber contains enzymes to which some people are particularly sensitive, which is why surgeons' gloves are now made from artificial latex. Nevertheless, the current fashion is to choose the product that claims to contain only 'natural' ingredients, and to pay a lot more for it. In fact, although it may claim to be 'natural', what this often means is that the ingredients can be found in Nature, not that they come from that source. Most come from chemical companies.

CHAPTER TWO

Vitality:
Food for Thought

PEOPLE IN THE DEVELOPED WORLD know they are most likely to die of either a heart condition or cancer, and consequently they are prepared to follow advice that might prevent this happening. Perhaps one day these diseases will become a thing of the past, or at least a rarity, like so many others that afflicted our parents and grandparents. Until that happy day, the best we can do is to heed the advice of those who advocate preventive measures, and these generally require us to change our lifestyles: give up smoking, take daily exercise, and eat more sensibly. For many people the easiest course to follow is a change of diet, and there are many nutritionists willing to offer sound advice, such as losing weight if you are overweight, and eating more fruit and vegetables, with five portions a day being the recommended amount (and not including potatoes).

Can the Grim Reaper's visit really be delayed by changes in our diet? The answer appears to be yes, and in this chapter we will look at three things that have had a lot of publicity in the past few years: something we are advised to eat less of, something we should be eating more of, and something they say we should avoid altogether. The first of these are fats, the second is vitamin C, and the one to avoid is nitrate. We will consider them mainly from the viewpoint of the chemist because, whatever else is said about these ingredients in our food, they are all chemical substances. The simple messages that most people have picked up about them are as follows: fats, and especially saturated fats and *trans* fats, cause heart disease; vitamin C is an antioxidant and can prevent cancer (perhaps even cure it); nitrates in drinking water can cause cancer. These messages are rather simplistic, however, and parts of them are surprisingly wrong.

Revelations about our food have resulted in improved human health and, in the first half of the twentieth century, there were many scientific discoveries about the things we eat, such as the need for various vitamins. In the second half of that century there were further revelations, although these were often less well researched. Thus, we had books advocating diets rich in zinc, magnesium, selenium, germanium, amino acids, and the ubiquitous fibre. Other books carried warnings of dangerous components such as unsaturated fats, sugar, salt, monosodium glutamate, pesticide residues, and cholesterol. Some of this advice was well worth taking but some was worthless.

Earlier generations were tied to the cycle of the seasons, with abundances of food following the harvest in autumn, and shortages in spring. Their diet may have been 'natural' and 'organic', but we know from their remains that they led unhealthy lives. Not surprisingly, they suffered deficiency diseases, such as scurvy through lack of vitamin C, and rickets through lack of vitamin D. The advances in food science in the twentieth century put an end to malnourishment of this kind, although cases of rickets are still being referred to dietitians for treatment even in Western societies.

The human body is basically a chemical processing plant, taking in raw materials and converting them to thousands of products, storing material that it cannot immediately use but which might be needed if supplies are short, and disposing of the things it cannot use along with its other waste products. The main raw materials the body factory needs are carbohydrate, fat, and protein, but they are not sufficient in themselves; the factory also requires moderate amounts of some metals, such as sodium, potassium, and calcium, and small amounts of many others, such as iron, zinc, and magnesium. Nor are these enough, and we also need a number of vitamins and trace elements. Finally, to ensure our intestines work smoothly we need to take in plenty of water and fibre, which consists of indigestible plant material such as cellulose.

It is pointless to make changes to your diet if you don't understand what's going on; it's a bit like trying to do a jigsaw puzzle in the dark. Yet the key to a successful diet is chemistry, and nothing illustrates this better than energy. Energy is not a nutrient as such, but it is the energy in food that our body needs to keep us warm, mobile, and thinking. Energy is released from food by reacting its components with oxygen, and we release the same amount of energy within our body as we would produce by setting fire to the food. The components that can release energy in this way are: carbohydrates (which release 4 calories per gram), proteins (which also release 4 calories per gram), alcohol (which releases 7 calories per gram), and fats and oils (which release 9 calories per gram).[7] The cellulose part of fibre is also a carbohydrate and is chemically like cotton or paper, which will also release 4 calories of energy per gram if burnt, but our body cannot release this energy because we don't have the enzymes capable of breaking it down into the glucose molecules of which it is composed.

Clarifying the fats

Fats and oils are described in various ways, as saturated, mono-unsaturated, polyunsaturated, omega-3, omega-6, *trans*, and CLA. All these terms relate to just one kind of chemical bond, the double bond. Understand this simple bond, and all becomes clear.

'Fat' is a dirty word, and as a way of describing someone it is little short of an insult. Even when we are talking about food, 'fat' still has undertones of disapproval because we know that eating too much fat can make us fat. But things don't stop there, and it is even claimed that some kinds of fat, namely saturated fats and *trans* fats, threaten our health.

Despite all the talk about dietary fats and what they do to us, there is little understanding of what they are chemically and how they behave in the body. Many people appear to think that the fat they eat immediately becomes part of their own body fat, whereas it has first to be digested in the gut before it can be absorbed into the body, which means breaking it down into its molecular components. These are just a collection of smaller molecules that the body can use in all kinds of ways, mainly to extract energy but also to make new molecules. But it's the amount of energy fats can release that is the problem. A body that finds itself with surplus energy food begins to make its own body fat, which it stores all over the body and especially in parts that seem particularly noticeable.

---

[7] A calorie is the heat required to raise 1 gram of water by 1 °C; this is too small to be useful when talking about the human body, so we talk about kilocalories instead, which is the heat required to raise 1 kilogram of water by 1 °C. The common word for kilocalories is just calorie, however, which is how it is used here.

The obvious way to avoid becoming fat is to eat less of the high-energy foods, and this can be done by eating more of those that contain only a little fat. Food manufacturers are only too willing to supply such foods, as the shelves of supermarkets show, and they even make fat-based spreads that are labelled as 'reduced calorie' and even 'low fat'. The reason is that they contain a lot of water, but they still look like fat because the water is kept suspended in the fat by adding emulsifying agents. Typical low-fat spreads contain around 40% fat, while very low-fat spreads have around 25%. Rather oddly, the producers of vegetable oils have enjoyed support-ive publicity for many years, even though these are full-fat products, although occasionally things have not gone smoothly, as the box shows.

The average person eats too much fatty food, consuming about 100 grams of fat a day and providing 900 calories, whereas they could man-age with only 10 grams, providing only 90 calories, although this might not guarantee an adequate intake of fat-soluble vitamins and essential oils. (Generally around 25 grams is required, and this amount would pro-vide 225 calories.) Our fat intake is likely to come in the following ways: meat products account for 25%; cakes, cookies, biscuits, and confec-tionery for 20%; spreads, 15%; milk and milk products such as cheese and yoghurts, 15%; vegetables, 10% (of which chips and crisps account for most); fish, 5%; eggs, 5%; the remaining 5% comes from things like nuts and sauces.

Nutritionists advise us to eat less fat, and clearly the easiest way to do this is to cut down on meat products, cakes, biscuits, and fried foods. Where possible we should eat fish in place of meat,[8] and use low-fat spreads. These sacrifices need not be great, and in terms of enjoyment you will lose little, but you might well lose some weight, and that in itself could well be the best way to lengthen your life and improve its quality. Yet we cannot exist on a diet entirely free from fat, because there are some vitamins that occur only in fats, namely the fat-soluble vitamins A, D, E, and K. Of the fats we eat, the aim should be to take in twice as much unsaturated as saturated fat. The foods highest in unsaturated fats are the vegetable oils, such as olive, sunflower, corn, rapeseed, soya, and

---

[8] Dietitians also say that it is best to eat oily fish, such as mackerel, herring, salmon, and trout. The minimum fish consumption suggested by dietary authorities is two por-tions of fish per week, one of which should be of an oily fish.

When margarine manufacturers ended up with egg on their faces

In 2001, the United Kingdom's Advertising Standards Agency (ASA), which deals with complaints about the accuracy of advertisements, reported on Benecol margarine spread. This had been launched in 1999 with the claim that it could reduce cholesterol levels in the blood by an average of 14%, and especially of the cholesterol associated with low-density lipoproteins (LDL), which is the dangerous type. Some thought this was worth paying a lot for, and Benecol was able to charge consumers more than three times the price of ordinary spreads.

Makers of competitor margarine, Flora Pro.active, complained to the ASA that Benecol advertisements were misleading. The ASA agreed. What the advertisements failed to mention was that the 14% reduction required a person to eat 32 grams of Benecol per day (most people eat only 20 grams per day of such spreads) and be in the age range 50–59. Tests had shown that people younger than this would experience much less of a cholesterol reduction. The ASA told Benecol to drop the claims.

But there is a twist to this tale. The ASA also passed judgement on Flora Pro.active and the claims its manufacturers were making. These said that eating 20 grams a day of their own (overpriced) product would reduce the dangerous cholesterol by 10 to 15% within three weeks. Although this was a more realistic claim, it too fell foul of the ASA who said that the people shown in the Flora Pro.active advertisements were clearly already following a healthy lifestyle, and the effect on their cholesterol levels would be much less, so those advertisements were also misleading.

And who had complained to the ASA about Flora Pro.active? You've guessed it: Benecol!

---

some nut oils (but not coconut oil, which is the most saturated fat of all). The foods highest in saturated fats are dairy products, suet, and beef fat.

Oils and fats are the same kinds of chemical; they are fatty acid derivatives of **glycerol** (see Glossary) and are known as triglycerides. Triglycerides are also called lipids, but most of us refer to them as fats or oils. In fact, whether it is solid fat or liquid oil depends on the triglyceride's melting point. The molecular structure of a triglyceride is rather like a capital letter 'E' with elongated horizontal strokes. The vertical of the E represents the glycerol molecule and the horizontal strokes are the **fatty acids** (see Glossary); it is these that are described as unsaturated, mono-unsaturated, polyunsaturated, *trans*, omega-3, omega-6, and conjugated linoleic acid (CLA).

The triglycerides in foods contain several types of fatty acid: for example, a hen's egg contains 11% fat, of which 3% is saturated, 4.5% is mono-unsaturated, and 3.5% is polyunsaturated. Of the polyunsaturated fatty acid component, 1.6% is omega-6 and 0.1% is omega-3. *Trans* fatty acids account for a mere 0.1%, and CLA for even less, and the fat make-up is irrespective of whether the eggs were laid by caged or free-range hens.

All the chemical terms used to describe the fatty acids relate to the concept of a carbon-to-carbon double bond; understand that and all becomes clear. But first let us look at the fatty acids with no double bonds, the saturated fatty acids.

Fatty acids consist of a chain of carbon atoms, with the carbon at one end of the chain being part of an acid group; this is a carboxylic acid group, CO_2H. In a saturated fatty acid, every other carbon along the chain has two atoms of hydrogen attached (shown as CH_2–CH_2 in chemical formulae, with a single dash showing a single bond). The chains almost always have an even number of carbon atoms, of which palmitic acid, with 16 carbons, and stearic acid, with 18, are the most common. The names of the fatty acids found in foods are as follows:

Saturated fatty acids in foods

No. of carbon atoms in chain	Name[9]
4	Butyric acid
6	Caproic acid
8	Caprylic acid
10	Capric acid
12	Lauric acid
14	Myristic acid
16	Palmitic acid
18	Stearic acid

[9] These are the common names for the acids, established by long usage, but the acids also have more systematic chemical names based on the number of carbon atoms in the chain; e.g. palmitic acid is hexadecanoic acid (from hexa = 6 and deca = 10), and stearic acid is octadecanoic acid (from octa = 8 and deca = 10).

There are saturated fatty acids with even longer chains, but these have no part to play in the chemistry of food, and there are some with an odd number of carbon atoms but these are rare, though not unknown. For example, there are traces of pentadecanoic acid, with 15 carbons, and heptadecanoic acid, with 17 carbons, in milk and in some plant oils, while valeric acid (5 carbons) and enanthic acid (7 carbons) are part of the characteristic odours of such foods as cheeses, where they are formed by micro-organisms.

Saturated fats are generally solids because such chains tend to arrange themselves neatly in line with one another, thereby packing together better as molecules, and it is this that raises the melting point. On the other hand, if one or more of the fatty acid chains are unsaturated, then packing becomes uneven, melting points decrease, and the result is an oil under normal temperature conditions.

Remove a hydrogen atom from each of two carbon atoms that are next to each other in a saturated fatty acid chain, and you end up with a carbon-to-carbon *double* bond; the chain becomes 'unsaturated' at that point because it has fewer hydrogen atoms than theoretically possible. Chemists show a double bond thus: $CH=CH$. The bonds that the two carbons once formed to separate hydrogen atoms now form a second bond to each other. The most common unsaturated fatty acids in foods are oleic acid and linoleic acid, both of which have 18 carbons atoms. They differ in that the former has one double bond, the latter has two; in other words, the one is mono-unsaturated (*mono* being the Greek word for '1'), the other is polyunsaturated (*poly* being the Greek word for 'many').

Mono-unsaturated fatty acids

Those found in foods are as follows, along with their omega designation that indicates where the double bond lies along the hydrocarbon chain.

Mono-unsaturated fatty acids in foods

Common name	No. of carbons	Omega labelling
Myristoleic acid	14	Omega-5
Palmitoleic acid	16	Omega-7
Oleic acid	18	Omega-9
Elaidic acid	18	Omega-9
Erucic acid	22	Omega-9

In the omega notation the carbon atoms are numbered starting at the carbon farthest from the acid group; in other words, the last carbon in the chain (omega is the last letter of the Greek alphabet). Thus, oleic acid, which is an omega-9 oil, and the most common, has its double bond between carbons 9 and 10. As mentioned earlier, there are also omega-3 and omega-6 fatty acids, but these are polyunsaturated and will be dealt with in more detail below.

In theory, the double bond of an 18-carbon-chain fatty acid, like oleic acid, could be between any pair of carbon atoms, but Nature produces only one variety, and that is omega-9 oleic acid, which occurs in things like olive oil (of which it makes up 75%), rapeseed oil (63%), and peanut oil (55%). Even lard contains 43% omega-9 oleic acid.

Not all mono-unsaturated oils are deemed to be healthy, and erucic acid is one that is to be avoided, although even this has its supporters. Rapeseed oil contains as much as 25% erucic acid and, when rats were fed high levels of this oil, they developed disorders of their lipid metabolism. The implication was that humans might also be affected, and an intensive plant-breeding programme was initiated to produce low-erucic acid oil. This was successful and rapeseed oil that is destined for human consumption contains less than 2% erucic acid; this oil is known as canola oil in North America. The name 'canola' is derived from *Canadian oil*, and it can have as little as 0.5% erucic acid. Rapeseed oil for industrial uses, such as lubrication and as a hydraulic fluid, still consists of the high-erucic acid variety, this being essential to its operation as a high-temperature lubricant.

Yet erucic acid may not be quite as dangerous to humans as the animal tests would suggest, mainly because we are unlikely to consume it at anywhere near the levels at which it was fed to those hapless rodents. Indeed, in some cases a high erucic acid diet has been used as a form of medical treatment, and it gained a high profile as a result of the popular film *Lorenzo's Oil*—see box.

In general, though, the trend is still towards producing rapeseed oil with no erucic acid, and this could be accomplished by genetic modification. The canola cultivars were developed for growing in Canada, yet in some countries of the world they did not flourish, and in India, where rapeseed is the second most important oilseed crop, scientists have adopted a different approach. In 1998, the Indian biotechnologists,

Lorenzo's Oil

The 1992 film of this name was directed by George Miller, and starred Nick Nolte, Susan Sarandon, and Peter Ustinov; Nolte and Sarandon were nominated for Academy Awards for their performances. It was based on a true story about a boy, Lorenzo Odone, who develops adrenoleukodystrophy (ALD) and becomes bed-ridden, blind, and spastic. ALD is a genetic disorder that prevents the body from correctly dealing with very long-chain fatty acids, and it results in damage to the myelin sheath that surrounds and insulates nerve cells in the brain, with the result that the brain cannot process information properly. The condition affects only boys, and the defective gene is passed on from the mother.

Lorenzo's parents, Augusto and Michaela Odone, refuse to accept that nothing can be done for him, and his mother, especially, devotes her life to his care and treatment, becoming an expert on the condition. One symptom of those suffering from ALD is a high level of very long-chain saturated fatty acids in the blood. Could these actually be the cause of the condition? His parents thought they were, and they believed that their son would benefit from being fed large amounts of an unsaturated, very long-chain fatty acid. For this purpose they chose erucic acid, and were able to source it from the chemical company Croda, based at Hull in England, one of whose chemists, Don Suddaby, took up the challenge of extracting this oil.

When Lorenzo was fed erucic acid, the amount of very long-chain saturated acids in his blood fell to normal levels. But it was a false dawn. Even as the film was released, there were those who had conducted tests on many other boys with ALD and who knew by then that the erucic acid treatment was ineffective. In true Hollywood fashion, the film ends with signs that Lorenzo might be responding to treatment and, by highlighting Lorenzo's condition, the film may have served to raise public awareness of a genetic disorder that affects 1 boy in 25,000. Sadly, the real-life Lorenzo made no dramatic recovery from ALD, and he remained bed-ridden and helpless into his twenties.

A. Agnihotri and N. Kaushik, genetically engineered a strain of rape that produces no erucic acid at all. Its oil content is mainly oleic and linoleic acids, with levels of these that are more than double those in normal rape.

Polyunsaturated fatty acids

If there is more than one double bond in a fatty acid chain, it is said to be polyunsaturated, and the two most common acids of this kind are linoleic acid, with 18 carbons and two double bonds (between the 6th and 7th carbons, and the 9th and 10th carbons), and linolenic acid, also with 18 carbons but with *three* double bonds (between the 3th and 4th, the 6th and 7th, and the 9th and 10th carbons). The first double bond along the chain is the most important one as far as our body is concerned, and thus nutritionists talk of omega-3 and omega-6 fatty acids, with linolenic acid being the former type and linoleic acid being the latter.[10] The common polyunsaturated fatty acids are as follows:

Polyunsaturated fatty acids in foods

Common name	No. of carbons	Number of double bonds in chain	Omega labelling
Linoleic	18	2	Omega-6
Linolenic	18	3	Omega-3
Arachidonic	20	4	Omega-6
EPA (eicosapentaenoic)	20	5	Omega-3
DHA (decosahexaenioic)	22	6	Omega-3

Polyunsaturated fats have an inbuilt disadvantage that comes with having double bonds. While saturated fats resist oxidation, the double bonds of unsaturated fats make them susceptible to oxidation. Not that the oxidation is a rapid process, but the threat is always there and, for this reason, precautions have to be taken to ensure that this does not occur, because it makes the oil or food go rancid. Oxidation can even produce a solid 'skin' on an oil, as used to occur when paints and varnishes were based on linseed oil, of which unsaturated fatty acids make up 95%.

The same oxidative processes occur in the body, and this might explain why animals that live a long time tend to have more saturated fats, because this reduces oxidative damage and stress. In 1998, Reinald Pamplona and colleagues at the University of Lleida, Spain, published

[10] Readers should note the close similarity of the words linoleic and linolenic, but these are very different fatty acids.

their researches into the fatty acids in the livers of animals. They found a relationship between the maximum lifespan of an animal and the amount of saturated fatty acid in the cells of its liver. The longer an animal lived, the more saturated were the fatty acids in its flesh. Humans are long-lived creatures and consequently a high level of saturated fats in our flesh is to be expected; it is not something that is intrinsically detrimental. Indeed, if this theory is correct, it is something to be thankful for.

Each triglyceride molecule contains three fatty acids, and these can be all the same or all different. In fact, things are more complicated than just being a pick-and-mix combination, and there can (in theory) be tens of thousands of different **triglycerides** (see Glossary). Nature is content with only a few of them; for example, olive oil is mainly oleic–oleic–oleic (O–O–O), although fifteen different triglycerides have been identified. Soya bean oil and sunflower oil, on the other hand, have very little oleic–oleic–oleic; their main oils are linoleic–linoleic–linoleic (L–L–L) and linoleic–linoleic–oleic (L–L–O). Altogether, soya bean oil has around fourteen different triglycerides, while sunflower oil has twenty or more.

The analysis of lard, which is produced from pig fat, shows that it is mostly made up from *unsaturated* fats, which is not how most people think of it, assuming that it must be mainly a saturated fat because it is solid and of animal origin. Lard is much used in baking. Its molecular composition shows that oleic–palmitic–oleic (O–P–O) accounts for 18%, followed by stearic–palmitic–oleic (S–P–O) at 13%, and O–O–O at 12%, with more than thirty other combinations accounting for the remainder.

Some triglycerides are predominantly made up of one variety and, in cocoa butter, it is S–O–P. This consistency is why chocolate melts over a relatively short temperature range, corresponding to that of the inside of the human mouth.

Essential fatty acids

Research on rats carried out early in the twentieth century eventually showed that there were two fatty acids that could not be eliminated from the diet, otherwise the animals suffered a whole raft of disorders. These fatty acids were necessary for the production of a group of key body chemicals known as prostaglandins, and they were deemed 'essential'; in other words, they had to be part of the diet because there was no way they could be synthesized within an animal's body. The human body can produce all the saturated and mono-unsaturated fatty acids that it needs from other food components in the diet, namely carbohydrates, alcohol, and even proteins. It cannot make omega-3 and omega-6 fatty acids, however, and yet we need these for some of our metabolic processes. In particular, we need them to generate arachidonic acid, which is the source of many other chemicals that are involved in the body's defences, such as blood clot formation, as well as being responsible for the local inflammation that alerts the body to tissue damage.

Humans, like other animals, can synthesize saturated fatty acids up to 18 carbons long, and can even make unsaturated fatty acids by removing hydrogen atoms from the saturated chains, but they can do this only at certain points along the chain. Thus, while it is possible to convert stearic acid to oleic acid, by removing hydrogen atoms at carbons 8 and 9, we cannot convert it to linoleic acid (omega-6) by removing hydrogen atoms at carbons 6 and 7, and so linoleic acid is one of the fatty acids deemed essential. The same is true for omega-3 fatty acids, and it means that we need an outside source of these if our body is to function properly.

The book, *The Omega 3 Phenomenon: The Nutrition Breakthrough of the '80s*, by Donald Rudin, Clara Felix, and Constance Schrader (published in 1987) was based on the theory that modern diets were deficient in essential fatty acids, the lack of which, they said, was the cause of many contemporary health problems. According to Rudin, fatty acids regulate virtually every function of the human body, omega-3 oils especially so. Were they to be restored to our diet, then a whole host of ailments would disappear: heart disease, arthritis, skin problems, allergic disorders, ageing, behaviour problems in children, mental illnesses such as schizophrenia and agoraphobia (fear of open places), diabetes and, of course, various forms of cancer. It seemed too good to be true—and it was. Nevertheless, the underlying message of the book was sound: in other words, essential fatty acids must always be an important part of our diet.

Which foods are rich in omega-6 and omega-3 oils? Fish oil is one answer and particularly that from herring, mackerel, salmon, and trout. The omega-3 fatty acids are produced in algae, and this passes along the marine food chain until it reaches fish, and ultimately humans. The Inuit of North America used to eat a diet that probably had the highest fat content of any human group, and yet they suffered hardly any heart disease due, it was said, to their high intake of essential fatty acids. Ergo, we should emulate them and supplement our diet with a source of omega-6 and omega-3 fatty acids, such as linoleic and linolenic, and this is easily done by taking cod liver oil, halibut liver oil, corn oil, or evening primrose oil.

Evening primrose oil comes from the seeds of *Oenothera biennis* and is a rich source of linolenic acid, containing about 10% of this fatty acid, which is even more abundant in the seed oils of the borage plant (starflower) accounting for around 20%. Evening primrose oil has been prescribed for the treatment of eczema—with 250 mg being a typical daily dose—but claims that it relieved the symptoms of post-menopausal stress and rheumatoid arthritis were not borne out when double blind tests were carried out. Claims that it is a preventive against heart disease, high blood pressure, and asthma appear equally misguided.

The recommended daily intake of essential fatty acids for an adult should be about four grams of omega-6 fatty acids a day and at least one gram of omega-3. Thankfully, they are plentiful in our diet. Both linoleic (omega-6) and linolenic (omega-3) acids are present in all kinds of foods, and are particularly abundant in fish oils and vegetable oils. Sardines, herring, mackerel, and salmon are rich sources, as are the vegetable oils from soya beans, rapeseed, and walnuts.

Human fat

We are fed on fats from the moment we are born, and our body produces *lipase* enzymes to digest the fatty acids in our food; to these it matters not which ones they are: all are broken down into smaller components. They are not immediately stored as part of our own fat, so that eating lots of mono-unsaturated fatty acids does not mean that these will be stored within our own fat deposits, thereby changing its composition. Human fat is made up as follows: oleic acid, 49%; palmitic acid, 27%; linoleic acid, 9%; palmitoleic acid, 8%, and stearic acid, 7%. This represents a ratio of saturated (palmitic, stearic) to unsaturated (oleic, linoleic, palmitoleic) of 34% to 66%; and the unsaturated fats consist of 57% mono-unsaturated (oleic, palmitoleic) and 9% polyunsaturated (linoleic). The omega-6 acids comprise 9%, while the omega-3s are absent.

At the start of our lives we have no control over our fat intake. Women with babies produce breast milk, which is a complete food in itself.

It provides around 750 calories per litre, and a baby generally needs about this volume per day. Its composition varies during the first two weeks, when the protein content falls by a half (from 23 grams per litre to 11 grams), while the carbohydrate content increases (from 57 grams of lactose per litre to 70 grams), as does the fat content (from 30 grams per litre to 45 grams). Indeed, the fat content increases as the baby is actually feeding, and may go up from as low as 10 grams per litre to 60 grams per litre at the end of the feed. The reason for this lies with the way the milk is produced. That from the front of the breast, known as the 'fore' milk, is lower in fat and higher in lactose and water, and this is what a baby drinks first. The 'hind' milk, which is produced by cells in the tissue at the back of the breast, is higher in fat and is drunk later.

The fat of human breast milk is 50% saturated, 50% unsaturated. The saturated fats are palmitic acid (26%), stearic acid (8%), myristic acid (8%), lauric acid (5%), and arachidic acid (1%), plus traces of other acids. The mono-unsaturated fats are oleic acid (35%) and palmitoleic acid (3%), while the polyunsaturated one is mainly the omega-6 linoleic acid (10%), with traces of others. The omega-3 acids are linolenic acid and arachidonic acid, and these account for 0.9% and 0.6%, respectively.

Formula milk for babies has been improved over the years with the addition of polyunsaturated fatty acids to make it more like that of breast milk. Research in the United Kingdom has even claimed to have found that children reared on such improved formulas were more intelligent than those raised on conventional formula feed. This seems improbable.

There are two types of triglyceride about which dietary advice has been given in recent years; these are *trans* fats, and CLA.

Trans fatty acids

Hydrogenation of vegetable oils will turn them into fats, but not quite the same type of fat that Nature prefers. Whether this new type of fat is harmful is still unresolved.

In the table of mono-unsaturated fatty acids on page 39 there are two acids, oleic and the much rarer elaidic, which appear to be the same, in that each has 18 carbon atoms and each has the double bond between atoms 9 and 10, so both are omega-9 acids. Nevertheless, they are different, and that difference arises because of the nature of the double bond itself: in oleic acid the bond is *cis* and in elaidic it is *trans*. On the other hand, all the polyunsaturated fatty acids in the table on page 42 are *cis* fatty acids.

In a *cis* double bond, two hydrogen atoms of the double bond, CH=CH, are facing each other and are on the same side of this double bond, rather like the ends of the letter C. In a *trans* double bond they are placed diagonally opposite across the bond, rather like the ends of the letter S.[11] *Cis* fatty acids are thought by some to be better for us than *trans* acids, although the evidence that *trans* acids are actually harmful still remains inconclusive. In the United States the daily intake of *trans* fatty acids is around 8 grams, and in Britain it is around 7 grams, of which 2 grams are from animal sources, the rest coming from modified plant oils.

Wherever there is a *cis* double bond along a fatty acid chain, it results in a bend at that point and the chain goes off at an angle. On the other hand, if there is a *trans* double bond in the chain it merely causes a slight kink in the chain, and the two longer halves of the chain continue to lie in the same direction. The upshot of this is that *cis* chains break up the molecular packing arrangement that can be achieved by saturated and *trans* saturated fatty acids.

Although oils can be used for frying and baking, fats are a little more versatile in that they can also be used for spreading. In addition, they are more convenient when it comes to packaging them, and when we want to use and serve them at table. Converting oils to fats can easily be done by reacting them with hydrogen gas at elevated temperatures and pressures, thereby converting double bonds to single bonds. This process of hydrogenation requires a nickel catalyst to make it work efficiently. (None of the nickel gets into the final product.) To what extent the fatty acids become saturated depends on how hard you want the final product to be; the more you saturate the chain, the higher the melting point of the product, and the more likely it is to be a hard fat. The addition of two

---

[11] Chemists also use a shorter notation, calling the *cis* arrangement z (from the German *zusammen*, meaning 'together') and the *trans* arrangement e (from the German *entgegen*, meaning 'opposed').

hydrogens to a double bond generally occurs in the chain with most double bonds, and at the double bond that is nearest the end of the chain. The extent of hydrogenation is controlled to ensure that fully saturated triglycerides are not produced.

What also occurs during hydrogenation, and of which the earlier manufacturers of margarine were unaware, is that *trans* double bonds are being formed in the fatty acid chains. The total amount of *trans* fatty acids can be as low as 5%, or less, in premium margarines, but as high as 40% or more in commercial frying fats. The upshot is that in Western countries most people have a daily intake of around five grams of *trans* acids, most of which comes from hydrogenated oils. Some of the *trans* acids, however, come from grass-feeding animals, such as cows and sheep, and the bacteria that they rely on for digestion also have the ability to produce *trans* fatty acids, and these pass into their flesh and thence into the human diet. Small amounts of *trans* fatty acids are even present in some plants, including peas and cabbage.

In the 1930s, a group led by the Canadian biochemist R. G. Sinclair began to wonder whether *trans* fatty acids were metabolized in the same way as *cis* fatty acids, and concluded that in general they were. But in the 1980s it was shown that they may affect certain enzyme systems differently, at least in experimental animals, and possibly in humans. Research on humans also showed that *trans* fatty acids increased the level of the low-density lipoprotein (LDL), which is where most of the cholesterol in the blood is to be found. Because of this, *trans* fatty acids were being blamed in the 1990s for causing heart disease, but some saw them as much more sinister and linked *trans* fatty acids to diabetes, breast cancer, and prostate cancer. Others went even further and said they were potentially damaging to the foetus (causing babies to be born underweight), and to breast-feeding babies (causing disruption of their endocrine system).

The best-known researcher into the health effects of *trans* fatty acids was Walter Willett of the Harvard School of Public Health in the United States. His work brought *trans* fats to the attention of the media and the general public. Starting in 1994, he and his co-workers found epidemiological evidence that linked *trans* fatty acids to heart disease in men and in women. Willet's opinions were widely disseminated, thanks in part to the *New England Journal of Medicine*, the influential medical journal that published a supportive editorial expressing the opinion that *trans* fatty acids in the diet definitely increased the risk of coronary heart disease. Willett's academic standing, and the seemingly independent endorsement of his work by this world-class periodical, brought the issue to the attention of the media, and *trans* fats became such an issue that the US Food and Drug Administration even suggested that foods should be labelled with their *trans* fatty acid content.

Not everyone agreed with what Willett was saying, and his findings prompted other researchers to investigate; they came to very different conclusions. For example, a study in Scotland, where the population is particularly prone to early death from heart disease, failed to find any link with *trans* fatty acids, and a similar study covering nine European countries and Israel came to the same conclusion. (Other work showed that our body could digest these fatty acids in the same way as *cis* fatty acids.) On the basis of such contradictory evidence, Willett's findings have been criticized as little more than 'junk science' by some, although this seems a little harsh. Undoubtedly Willett was well intentioned, if perhaps a little hasty, in coming to the conclusions that he did.

Part of the problem in assessing *trans* fatty acids is that it is very difficult to analyse a food for its *trans* fatty acid component, generally requiring a combination of two analytical techniques: infrared spectroscopy and gas chromatography. In theory, either method by itself should be capable of measuring *trans* fatty acids, but a study by the Laboratory of the Government Chemist in the United Kingdom showed that when the methods were used in isolation they gave results that differed by as much as 20%.

Testing the effects of *trans* fatty acids is generally done with the three acids which have 18 carbon atoms in their chain: stearic acid, which is saturated; oleic acid, which has a *cis* double bond half-way along its chain; and elaidic acid, which has a *trans* double bond half-way along its chain. Elaidic acid is the main component of the *trans* fatty acids produced by hydrogenation of vegetable oils.

In 1990, Ronald Mensink and Martijn Katan, of the Agricultural University in Wageningen in the Netherlands, fed these three fats to three different groups of people, who agreed to eat exactly the same diet except for the inclusion of the different fatty acids. At the end of the experiment, it was found that those given the *trans* fatty acid had higher levels of low-density lipoprotein (LDL), plus lower levels of the high-density lipoprotein (HDL), the 'good' form of cholesterol. The number of people taking part in this experiment was small, so perhaps we should not read too much into the results.

Meanwhile, at Harvard University a group was conducting an eight-year analysis of 90,000 female nurses, and its results were published in 1993. In this study the intake of *trans* fatty acids was based on what the nurses said they ate, but the outcome was that there appeared to be only a slightly increased risk of heart disease among those with the highest intake of *trans* fatty acids.

Further reports on *trans* acids appeared throughout the 1990s, with perhaps the most prestigious being a joint one issued in 1996 by the American Institute of Nutrition and the American Society for Clinical Nutrition. This said that, at current levels of *trans* fatty acid intake, there was no risk to health. This was in line with the observation that, while *trans* fatty acid intake among the population of the United States has remained at around eight grams per person per day for more than half a century, the death rate from heart disease has fallen by more than a third. In 1999, John Stanley of Oxford University came to the same conclusion, that *trans* fatty acids were not the cause of heart disease, and pointed out the defects in some of the earlier studies that had found a link, although he sympathized with the investigators because of the difficulties they faced in trying to assess the dietary intake of *trans* fatty acids.

The British Nutrition Foundation also reported that 'animal studies have not indicated adverse effects of *trans* fatty acids on longevity, reproductive performance, or growth. In addition, they have not indicated teratogenic, carcinogenic, or mutagenic[12] potential, nor revealed any specific abnormalities in organs examined.' Nor did the report find any real evidence of *trans* fatty acids causing heart disease, which was the original cause for concern. Perhaps the later findings, that there was no link, were not unreasonable, because *trans* fatty acids are not unnatural chemicals formed during food processing; they also occur in Nature and have been eaten by humans since we first started herding sheep and eating their meat.

[12] Teratogenic means causing a foetus to be deformed; carcinogenic means cancer-causing; mutagenic means causing something like chromosome damage that may lead to a mutation.

Lamb fat contains 5 % of *trans* fatty acids, which might suggest that the populations of Middle Eastern countries, where such meat is widely eaten, would also show higher incidences of heart disease compared with countries where beef or pork are the norm. There is no evidence that this is so. Other animal fats have smaller *trans* components; beef fat contains only 2% *trans* fatty acids, while pork fat has less than 0.5%. Suet,[13] which is the hard fat found around the loins in beef and mutton, contains almost 6% *trans* fatty acids, although the suet substitute derived from vegetable oils contains twice as much. Butter contains around 3% *trans* fatty acids, soft margarines have about twice as much, and hard margarines twice as much again, whereas the vegetable oils from which the last two are made contain only trace amounts, less than 0.1%.

The amount of *trans* fatty acids in spreads and in the fats used in baking has been significantly reduced over the years as the companies that produce hydrogenated oils have researched the chemistry behind their formation and discovered means of reducing the amount formed. At one time, in some processed foods such as pastries, French fries, and fried chicken, *trans* fatty acids accounted for a third of their fat content. The manufacturers of soft margarines, however, have reduced the *trans* fatty acid content of these from around 10% in the mid-1990s to less than 1% today.

Conjugated fatty acids

These are relatively rare, but one in particular, conjugated linoleic acid (CLA), is being hailed as a dietary health benefit, and there is some reason to believe that it might even ward off breast cancer.

If two double bonds along a chain of 18 carbon atoms are on adjacent pairs of carbon atoms, say between atoms 6 and 7 and atoms 8 and 9, then the two bonds influence each other to a certain extent and also cause the single bond that connects them (in this case between atoms 7

[13] Suet is the 'meat' of mincemeat and mince pies. It is also the part of these animals that once yielded the tallow for candles and soap. Suet is 90% fat, of which half is saturated, half unsaturated, with the majority of the latter being mono-unsaturated.

and 8) to take on some double bond character as well.[14] The two double bonds are said to be *conjugated*. Such conjugated fatty acids are present in the fats of cows and sheep, and they are produced in the stomachs of these animals by anaerobic bacteria that live there, and on which the animal relies to help it digest its food.

Normal linoleic acid has two double bonds, the omega-6 bond between atoms 6 and 7, and a second double bond between atoms 9 and 10; in other words, there are two single bonds separating the two double bonds, and so there is no conjugation between them, and both are *cis* double bonds. In CLA, however, the first of the double bonds is a *trans* double bond, while the second is *cis*. There is also another form of conjugated linoleic acid, in which the double bonds are between atoms 6 and 7 and 8 and 9; the former is *cis* and the latter *trans* but, in most natural products, there is less of this.

The bacteria are able to convert linoleic acid to stearic acid by adding four atoms of hydrogen to the molecule. They do this in a rather complicated way that involves the intermediate formation of CLA, however, which then ends up in the foods derived from these animals, namely lamb, beef, milk, cream, butter, yoghurt, and cheese. Moreover, the amount of CLA depends on its source; tests in Scotland have shown that cows eating fresh grass produced twice as much as those fed on silage. The level of CLA is not high, but it may be more important than we imagine. Lamb fat can have 1.2% CLA, beef fat 0.6%, the fat in whole milk 1%, cream 0.8%, butter 0.9%, and the fat in cheese as much as 1.7%. The average person takes in about 0.4 grams (400 mg) of CLA a day.

The CLA story began with the work of Michael Pariza, at the Food Research Institute of the University of Wisconsin-Madison; he was investigating the effects of heat on minced beef, and discovered that it contained something that protected genes against mutation. Not only that, but raw beef contained the same agent as well. That was as long ago as 1979. Further research in the early 1980s by Pariza and his colleagues revealed that the active agent was CLA.

CLA is one of the most potent preventers of cancer—at least in rats.

---

[14] If you number the carbon atoms from the other end of such a chain, as food chemists do, then these double bonds are between atoms 9 and 10, and 11 and 12. The first of these is a *cis* double bond, the second is a *trans* double bond. CLA is sometimes referred to as *cis9,trans11*.

Exactly how it works is still unclear, or at which step it acts in the cancer process: that is, at initiation, promotion, or progression, or during the spread of the disease to other body organs. It may protect all along the line. It has also been shown to protect rabbits and hamsters against narrowing of the arteries, and to boost the immune system of mice, rats, and chickens. It is even a growth factor for young rats, and can stimulate the body to release energy from its fat stores. High intakes of CLA are reputed to offer protection, because CLA is chosen preferentially by certain enzymes in place of linoleic acid, which has been linked to an increased risk of cancer in rodents. Not only that, but studies in Norway also found that obese people put on a high CLA diet (more than 3 grams per day) actually lost weight.

CLA can be manufactured as a dietary supplement by treating the oils that contain a lot of linoleic acid, such as sunflower oil, with an alkali and, while the process converts this mainly to CLA, it also forms other conjugated acids with the two double bonds at various positions along the carbon chain. Other tests confirmed that all types of CLA could be digested by test animals and used as a source of food energy, with no observable difference from the normal types of dietary fats and oils. It is, however, probably somewhat premature to act on this research and move back to eating lots of dairy fats such as cream, butter, and full-fat cheese, especially if these displace low-calorie foods, such as fruit and vegetables, from the diet.

Vitamin C

Without vitamin C we would suffer from scurvy, as many did in years gone by. With the right foods we can get all the vitamin C we need for a healthy life. With mega-doses of vitamin C, some believe that we can avoid cancer. That was the theory of someone who won two Nobel Prizes—but it was probably wrong.

Dietary deficiency diseases are rare because sensible eating provides all that our body needs; even a diet of junk food is unlikely to cause them. By definition, a deficiency disease is the lack of a minor, but essential, nutrient, and this will invariably be one that is to be found in relatively few foods.

Iodine is a good example, and the lack of this causes the deficiency disease, goitre, with its characteristic swollen neck. It is prevalent in regions where the iodine content of the soil is low, and the disease has been more or less eradicated by adding small amounts of potassium iodide to common salt. Rickets is a childhood disease caused by lack of vitamin D and calcium, and this results in the deformity known as bow legs. Vitamin B1 deficiency leads to beriberi, which causes inflammation of the nerves and is likely to occur in times of famine. The best-known deficiency disease is scurvy, now thankfully very rare, but it afflicted former generations and was caused by a lack of vitamin C.

We need about 2000 mg of vitamin C in our body for its optimum performance, and we can absorb up to 500 mg per day although we may not need to use it; and the excess is excreted by the kidneys. In theory, we can get by on a basic minimum of vitamin C of only 10 mg per day, which is just enough to prevent scurvy, but there is now every reason to believe that ten times this amount will provide long-term benefits, and there are some who take a hundred times this as a daily dose, in the belief that it will prevent ageing, heart disease, and even cancer.

Vitamin C is needed to make hormones, and the highest concentrations are found in the adrenal and pituitary glands. It also protects cells against oxidation, and every cell in our body has to contend with thousands of damaging, oxidizing, **free radicals** (see Glossary) every day. Most of those free radicals are by-products from natural processes, but they are neutralized by vitamin C. Vitamin C is involved in the body's absorption of iron and in our ability to tolerate cold. It is essential for metabolizing the amino acids tryptophan, phenylalanine, and tyrosine, and for synthesizing polysaccharides and collagen. Vitamin C is needed for making cartilage, dentine, bone, and teeth. Vitamin C also protects the body when it is under stress and when exposed to ionizing radiation. Severe injury increases the body's utilization of vitamin C because it has a part to play in the growth of new tissue, so it is essential in wound healing and bone formation. All in all, vitamin C is truly remarkable, and its lack is truly disastrous.

Almost all animals make their own vitamin C, the exceptions being fish, bats, beetles, guinea pigs—and humans. Twenty-five million years ago our primate ancestors lost a gene for producing the enzyme *L-gulono-lactone oxidase*, without which we cannot make vitamin C. This channelled all who followed down that evolutionary branch to a mainly vegetable diet to get their daily dose of vitamin C; that included humans, although it is possible to get just enough from fresh meat, as we shall see.

Relatively few foods contain vitamin C, and that which is there we often destroy by cooking because it is sensitive to heat, or by long storage because it is also sensitive to oxygen. After boiling cabbage for ten minutes, a quarter of the vitamin C has been destroyed and most of what remains is in the water.

In the United Kingdom, the recommended daily amount (RDA)or reference nutrient intake (RNI)[15] of vitamin C is 40 mg, in the United States it is 60 mg, while in Russia it is 90 mg—but all may be on the low side. Sometimes we need much more than at other times—such as during pregnancy, when breast feeding, when our body has been injured, in old age, and when we are ill. Balz Frei, of the University of California at Berkeley, reported in 1989 that when all the vitamin C in blood has been used up oxidation damage to fats and proteins occurs, and suggested a minimum daily intake of 150 mg. Later research by Bruce Ames, also at Berkeley, suggests that even this may be too low. Ames believes that vitamin C has a special role to play in protecting sperm because of its relatively high level in semen, and showed that the more vitamin C there was, the less was the oxidation damage to sperm DNA. He concluded that men need at least 250 mg a day if they are fully to protect their sperm. (There is more about male fertility in Chapter 3.)

As far as humans are concerned, the problem is that the most common and tasty foods contain no vitamin C at all, and this goes for butter, cheese, eggs, margarine, meat, fish, poultry, bread, cakes, biscuits, chocolate, cereals, pasta, rice, nuts, and beans. Fruit and vegetables contain some, and a few have a great deal. The humble potato is a good source, and it was mainly this vegetable that was responsible for the disappearance of scurvy in the nineteenth century.

[15] Once referred to as the RDA, this acronym has been superseded by the more scientific RNI. The reason for the change was that the RDA was being interpreted as the minimum needed for a healthy life whereas, for many food components, the vast majority of the population takes in substantially more than this. The RNI represents the best estimate of the amount that is needed and, by avoiding the word 'recommended', it does not imply a basic minimum.

We need to be aware of foods rich in vitamin C and make sure we eat some every day. The following are particularly good natural sources:

Glass of blackcurrant juice	95 mg
An orange	90 mg
Dish of strawberries	60 mg
Portion of Brussels sprouts	45 mg
Portion of broccoli	30 mg
Portion of freshly made chips[16]	30 mg
Half a grapefruit	30 mg
Slice of lemon	20 mg
A tomato	15 mg

Other foods with at least 50 mg of vitamin C per 100 grams are red and green peppers, cabbage, and spinach, and though peas fall short of this target, they are eaten in large quantities and so constitute a major source of vitamin C. All fruits contain some vitamin C, and mangoes and honeydew melons are particularly rich in it. Human milk contains 50–70 mg vitamin C per litre, and a breast-fed baby is almost never at risk, given that it consumes an average of around 750 ml per day, which provides around 50 mg. On the other hand, fresh cow's milk contains only about 10 mg per litre, and this varies depending on the time of year, being lowest in early spring.

If people heed the dietary advice to eat five portions of fruit or vegetables every day then they are unlikely ever to need to take vitamin C as a dietary supplement. In any event, it is now added to many foods and drinks, especially processed foods in which the processing might well have removed what little natural vitamin C was present. Despite all these dietary sources, it may still be advisable if you eat a lot of fast food to take a daily dose of vitamin C in a tablet form, and this can provide up to 1000 mg.

Vitamin C is put in flour as an improving agent because it makes dough more elastic and leads to larger loaves of bread; but it does not survive the baking. It is used in fish farms to prevent broken-back syndrome, which is caused by vitamin-C deficiency. It is added to sliced fruits and vegetables to prevent browning, and to some foods as a preser-

---

[16] As in fish and chips, not as in French fries, which are made from reconstituted potato.

vative because of its antioxidant properties. When used as a food additive in Europe, it even has its own additive number: E300.

Plants make vitamin C for the same reason that most animals do, which is to protect them against oxidative damage, in this case by the ozone of the air as well as the free-radical by-products of photosynthesis. Plants also need it, the better to resist environmental stresses such as drought, cold, and air pollution. They require vitamin C for their own growth processes, and those with leaves that live a long time, such as pines with their needles, particularly need vitamin C to protect them.

Until recently, scientists were puzzled as to how plants actually made vitamin C. The problem was solved only in 1998 when Nicholas Smirnoff and co-workers at the Department of Biological Science at Exeter University, England, came up with the answer. They made the vitamin from glucose, via another sugar called galactose, and they showed that if leaves were fed this second chemical directly, they could produce vitamin C very rapidly. The Exeter group was even able to isolate the enzyme responsible for this conversion, *L-galactose dehydrogenase*.

Join the navy and see no C

We may no longer need to worry about scurvy, but it was an ever-present threat to previous generations, especially among those with limited access to fresh fruit and vegetables, such as seamen on long voyages, and city dwellers. Indeed, everyone living in northern countries was at risk at the end of winter when fresh foods were no longer available.

No doubt cases of scurvy occurred in the ancient world, but the disease appeared in large numbers only when European sailors began to make long voyages by sea. When the Portuguese navigator Vasco da Gama (c.1460–1524) first sailed round the Cape of Good Hope in 1497, he lost 100 men out of a total contingent of 160 to scurvy. Ferdinand Magellan (c.1480–1521), who sailed round the tip of South America in 1520, likewise found his men afflicted. The Frenchman Jacques Cartier (1491–1557), who set out to explore the regions of north-west Newfoundland in 1535, had to contend with an outbreak of scurvy but luckily help was at hand. He learned from the native people that the condition would respond to a decoction of leaves, and it did. Within a week the outbreak was over and his men were on the mend. What the leaves were cannot now be known—he referred to the tree by the native American name of

Hanneda—but stirring young pine needles in hot water produces a drink that is rich in vitamin C, and this may have been what was used.

By the end of the sixteenth century, scurvy was a recognized medical condition, and various herbal remedies were being proposed to treat it. John Gerard (1545–1612), in his 'Herball', published in 1597, suggested winter cress and scurvygrass (a cress-like plant that grows near the sea). Fleet commander James Lancaster, who was in charge of an expedition to the East Indies in February 1600, took oranges and lemons with him and, while there were many cases of scurvy on board the other ships, there was none on his, where he had insisted that every man be given a ration of three spoonfuls of lemon juice each morning. The East India Company became a firm believer in citrus fruits, and the company surgeon, John Woodall (1556–1643), extolled their virtues in his book *The Surgeon's Mate*, published in 1612. This treatment was effective, but the benefit was attributed to the acidity of the drink rather than to the source from which it came. Because fruit drinks were not easy to preserve in hot climates, alternative acidic drinks were used instead, such as vinegar or a few drops of oil of vitriol in water (sulfuric acid). This misplaced faith in sulfuric acid as a remedy continued for a hundred years.

In the advanced stage of scurvy the victim had swollen limbs, especially the legs and feet, which were often dark and discoloured as if badly bruised; this was caused by haemorrhages under the skin. The breath was foul-smelling because of ulcerated and bleeding gums, and often the teeth were loose. Convicts kept on hulks and those transported to the colonies were very prone to scurvy, made worse by harsh punishments.

Some doctors came up with advice that would eventually end the threat of scurvy: in 1734, a Dr Johannes Bachstrom of Leyden in Holland advised ships' victuallers to provision ships with vegetables as a means of preventing scurvy at sea. Sadly, his advice was not heeded, although it would have saved many lives. A Dr Mead, on the other hand, simply said that patients should breathe the vapour of cold earth, again advice that was not heeded, but that advice, had it been acted upon, would have saved no one.

James Lind (1716–94) wrote the classic work, *A Treatise of the Scurvy*, in 1753, and distinguished scurvy by three types of outbreak. The first, and most serious, was when 'it rages with great and diffusive virulence . . . as an epidemic or universal calamity', such as seamen experienced on long voyages, or among people in cities under siege, or even among whole populations, such as the widespread outbreak in Holland in 1562. The second type was less serious and was endemic in a population, in other words, there are always a few cases and the population is never completely free from the disease. Lind thought that people living in the far north— such as those in Iceland, Greenland, Scandinavia, and northern Russia —were most vulnerable. The third, and least serious, was the occasional person who went down with the disease, as happened in countries such as Britain, where it occurred mainly in London.

Lind was a great advocate of citrus fruits as preventives, and he based his claims on an experiment he carried out on twelve patients with scurvy aboard the ship *Salisbury*. In May 1747, he divided the men into six pairs and gave each pair a different remedy: cider, vitriol, vinegar, seawater, a patent medicine, or citrus fruit. (The patent remedy consisted of garlic, mustard-seed, horseradish, and myrrh.) The citrus fruit treatment consisted of two oranges and one lemon every day, and the two men given these were cured within a week, while the rest languished with the disease. Sadly, Lind's findings were not immediately acted upon, and in the Seven Years War (1756–63) between Britain and France more than 100,000 sailors in the Royal Navy were to die from scurvy. In 1781, 1600 died from scurvy in a British fleet of 12,000 men .

Some sea captains began to take notice of Lind's findings, however. For example, the English navigator Captain James Cook (1728–79) took sauerkraut on his three-year voyage around the world (1772–75), and he lost not a single man to the disease. But, while this preserved vegetable would have provided some protection, it was his policy of taking on board large quantities of fresh fruit and vegetables whenever he could that provided most protection.

Progress in eliminating the disease was painfully slow, and even measures that might have proved successful were negated by ignorance. In 1793, the Viennese physician Mertens observed that cooked vegetables were inferior to fresh ones in preventing scurvy, but fresh vegetables were not really an option for long journeys. Sailors and travellers lived mainly on oatmeal porridge, bacon, biscuits, cheese, and salted meat, none of which contained any vitamin C. Yet clearly something had to be done, and in the mid-1790s the British Admiralty finally ruled that every sailor on active service in the Royal Navy should be issued with a daily ration of lime juice. This would at least have prevented scurvy, even though this citrus fruit has less vitamin C than either oranges or lemons, but what little it had was considerably reduced when the juice was boiled to preserve it. Not surprisingly, scurvy continued to bedevil His Majesty's ships, albeit not as extensively as previously.

The Shipping Act of 1845 laid down that ships that were to be away from port for more than ten days must have anti-scurvy items such as lime or lemon juice as part of the seamen's rations (an ounce a day was prescribed, which would contain 5 mg of vitamin C if it were lime juice, and 10 mg if it were lemon juice); lime juice preserved with rum was the preferred issue. British sailors became known in the United States as 'limeys' as a result of these Board of Trade regulations. In 1927, the Board authorized ships' masters to issue half an ounce of concentrated orange juice a day, which would provide 30 mg of vitamin C. This ensured there would be no more scurvy among British sailors, but soldiers had not been so lucky.

Scurvy appeared in the British Army in the Crimean War of 1854–56, and in World War I in 1915. In the earlier war the disease quickly appeared, and by October 1854 it was ravaging the ranks. The Commander-in-Chief, Lord Raglan (1788–1855), tried to obtain fresh vegetables locally, but without success, so he wrote to the War Office in London for supplies of lime juice. Two thousand gallons were sent and reached the base at Balaclava in December 1854 but Raglan was a sick man and died in June 1855, and so the lime juice languished, forgotten, until the end of the war. The British hospital reformer Florence Nightingale (1820–1910), who went to the Crimea, noted that scurvy caused more casualties during that war than the fighting. She reported that, of the 1200 men who arrived sick at Scutari in January 1855, around 1000 had scurvy. Nor did the allied French Army fare any better, losing an estimated 20,000 to the disease.

Around this time, too, there was an outbreak of scurvy in the United States during the California gold rush of 1849. To reach the fabled wealth men either travelled overland or by boat via Cape Horn, living mainly on a diet of flour, biscuits, salt pork, and beef. It has been estimated that around 10,000 prospectors died of the disease.

By the early twentieth century, scurvy was rarely encountered in the armed forces, but in December 1915 it struck the British Army yet again.

General Townshend and the forces of the British Empire were in retreat after attempting to capture Ctesiphon, a Turkish stronghold on the River Tigris about 32 kilometres south-east of Baghdad. Townshend decided to make a stand at Kut-el-Amara, where the pursuing Turks soon besieged him. On Christmas Day strict rationing began, and as the weeks went by the rations were cut until, by the first week of March, the issue for British troops was 20 ounces of horsemeat and 10 ounces of bread a day; for Indian troops it was 10 ounces of barley flour, 4 ounces of whole barley, a little ghee butter, and a few dates. Both diets resulted in two kinds of deficiency disease: the British troops began to go down with beriberi, through lack of vitamin B1, and the Indians with scurvy through lack of vitamin C.

Horsemeat contains a little vitamin C, enough to prevent scurvy among the British, while whole barley contains B1, enough to prevent beriberi among the Indian troops, which explains the exclusive nature of the two outbreaks of deficiency disease. When Townshend finally surrendered at the end of April, there were 1100 cases of severe scurvy and 150 of beriberi out of his force of 9000 men. Scurvy was to bedevil the forces of the British Empire in Mesopotamia, as Iraq was then called, throughout World War I (over 10,000 men were afflicted with it at one time or another), and even lime juice could not entirely prevent it. This led to a reassessment of this juice and its replacement with lemon juice, which became standard issue to Empire forces in the East.

The discovery of vitamin C

Over the centuries there had been much speculation as to the cause of scurvy and suggestions as to how it might be prevented and treated. Things that were blamed were rancid butter, copper pans, rum, sugar, tobacco, dampness, cold weather, sea air, hereditary factors, low morale, lack of fruit, infection, too little exercise, and the onset of spring. Some of these suggestions were spot on, while others were indirectly linked to the real cause. For example, the end of winter and start of spring came at the end of several months without fresh fruit and vegetables, so naturally the incidence of scurvy was likely to reach its peak around then. The use of copper pans for cooking was also near the mark, because this metal acts to catalyse the reaction of vitamin C and oxygen, thereby rendering it useless.

The first piece of scientific evidence as to the true cause of scurvy came in 1907 when the Norwegian government funded some research into beriberi. Two doctors, Axel Holst and Theodor Frölich, began a series of experiments with guinea-pigs in which they tried to induce this condition by feeding the animals restricted diets. The result was that some of the guinea-pigs went down, not with beriberi, but with scurvy, showing that this was indeed a deficiency disease. As was then common in laboratory testing, they were working with an animal that also lacked the gene to make vitamin C.

Next to investigate the disease was Harriette Chick at the Lister Institute, London. In 1918, Chick identified various foods that would prevent scurvy in guinea-pigs. A year later, Jack Drummond called the anti-scurvy agent 'vitamin C', although what exactly it was remained a mystery. The chemical was first isolated from paprika in 1928 by the Hungarian biochemist Albert Szent-Györgyi[17] (1893–1986), who wanted to call it 'ignose' from the Latin word *ignorare*, meaning 'not to know', because he didn't know what it was. The editor of the *Biochemical Journal*, to which he had sent his paper, was not amused and rejected his choice of name on the grounds that they did not publish jokes.

Szent-Györgyi then came up with an even better name: 'godnose'. That, too, failed to crack a smile, and in the end he called it hexuronic acid and showed its chemical formula to be $C_6H_8O_6$. Although in his paper he hinted that this might be the agent that prevented scurvy, he could not prove it was. It was Charles King, of the University of Pittsburgh in the United States, who showed that hexuronic acid, which he had extracted from cabbage and lemon juice, was, in fact, vitamin C. It was Szent-Györgyi, however, who was awarded the 1937 Nobel Prize in Physiology or Medicine for this discovery, which he had made while working at the Mayo Clinic, Rochester, Minnesota. Szent-Györgyi had returned to Hungary in 1931 and was Professor of Medical Chemistry at the University of Szeged, where he remained until 1945. He emigrated to the United States in 1947 to take up the post of Director of the Institute for Muscle Research, Woods Hole, Massachusetts.

The making of vitamin C and massive profits

In 1933, Norman Haworth was Professor of Organic Chemistry at Birmingham University, England, and was a leading expert on sugars. He

[17] Pronounced as in 'Saint-George'.

was sent a sample of vitamin C, and his group succeeded in deducing its molecular structure. They also confirmed that their analysis was correct by synthesizing it chemically in the laboratory. It was at about this time that vitamin C took on the name ascorbic acid from the Greek words for 'no scurvy'. Haworth shared the Nobel Prize for Chemistry in 1937 for this work.[18]

Ascorbic acid has a five-membered ring of atoms consisting of one oxygen and four carbons, two of which are connected by a double bond (which is why vitamin C is prone to oxidation by the oxygen of the air, but also partly why it makes a good antioxidant). Once the molecular structure of vitamin C was known, it became clear that it could be made from common sugars, and especially glucose. Indeed, in 1933, Tadeus Reichstein was able to make it synthetically using glucose as the starting material. Chemical companies soon began to manufacture vitamin C on a larger scale, the first being the Swiss pharmaceutical company Roche, which began production in 1934. Today, more than 50,000 tonnes are made world-wide each year, a lot of this at a plant situated at Dalry in north Ayrshire, Scotland, which exports 90% of its production.

The process for making vitamin C involves several stages, the first of which is the conversion of glucose to sorbitol by reacting it with hydrogen. This is then fermented with *Acetobacter suboxydans*, which converts it to sorbose. This is oxidized by potassium permanganate, or other oxidizing agents, to a derivative of gulonic acid which is easily converted to ascorbic acid by treatment with hydrochloric acid. In 1985, a simpler two-step process was devised by the biotechnology company Genetech, in which a genetically modified bacterium was designed to convert glucose directly to gulonic acid.

The profits from manufacturing vitamin C, and other vitamins, were to become truly enormous, chiefly because the leading manufacturers operated an illegal cartel to keep the price artificially high. The cartel of Swiss, French, German, American, and Japanese companies was set up in 1989, with Hoffman-La Roche the leading partner. In the late 1990s, its activities were exposed and court cases followed in the United States and Europe. The members of the cartel agreed to pay compensation of $1.2 *billion* in the United States, of which Hoffman-La Roche paid half as

[18] He shared the prize with the Swiss chemist, Paul Karrer (1889–1971,) who had deduced the chemical structures of vitamins E, K, and B2.

the ringleader of the cartel. Meanwhile, in Europe the European Commission levied fines of €860 million on the companies, again with Hoffman-La Roche paying most at €460 million. The other guilty parties were BASF, Aventis, Solvay, Merck, and the Japanese companies Daiichi Pharmaceutical and Eisai. The fines were levied as a percentage of a company's annual turnovers. Nor surprisingly, the revenues from vitamin C manufacture slumped and, in Europe, they had fallen from €250 million to less than €120 million by 1998. (Roche sold its vitamin-production facilities to the Dutch chemical company, DSM, for €1.9 *billion* in 2003.)

Vitamin C, the common cold, and cancer

Although Linus Pauling (see box) championed the benefits of taking mega-doses of vitamin C, he was not the originator of this controversial treatment; Irwin Stone was. His book *The Healing Factor*, published in 1974, was based on a belief that ascorbic acid had a much bigger role to play in the body than had previously been suspected. Little notice was taken of Stone's opinions until, one day, he and Pauling got in to the same lift and started talking. It is said that by the time they emerged Pauling was a convert to the theory that ascorbic acid could protect the human body against almost any illness.

From that time onwards, Pauling carried the banner for vitamin C, advocating doses of at least 1000 mg a day for ailments ranging from the relatively minor, such as the common cold, to the deadly serious, such as cancer. He set up the Linus Pauling Institute of Science and Medicine in the 1970s to back up the claims he was making. Nevertheless, the medical establishment remained sceptical. Today, however, there seems to be more support for his ideas, because we appreciate the damage that free radicals can cause and that antioxidants, such as vitamin C, are essential in combating them.[19]

The popular notion that vitamin C is a good treatment for the common cold received support in 1987 when Elliot Dick, head of the Virus Research Laboratory at the University of Wisconsin, showed that it both relieved symptoms and reduced transmission of the virus. Researchers

[19] There are still a large number of 'believers' who regard vitamin C as a wonderful cure-all, and they have a web site which is a useful source of information, albeit somewhat biased: www.vitamincfoundation.org/

Linus Pauling (1901–94)

Linus Pauling won two Nobel Prizes: the Chemistry Prize in 1954 and the Peace Prize in 1962. Although no official reason was given for awarding the Peace Prize to Pauling, it is generally accepted that it was for his work in alerting the world to the dangers of testing nuclear weapons in the atmosphere. His book *No More War!* (1958), and the petition he presented to the United Nations, signed by 11,021 scientists from around the world, were instrumental in bringing about the Test Ban Treaty, which was signed the very day he was presented with the Peace Prize.

Pauling's Chemistry Prize was awarded for his ground-breaking work on chemical bonding and molecular structure. His greatest achievement was his insight into the nature of how atoms come together to form molecules, and how their structure could be explained. His textbook *The Nature of the Chemical Bond* was first published in 1939, and is regarded as a classic of its kind.

Pauling saw that chemistry was the key to understanding biological molecules, such as antibodies, haemoglobin, and proteins, and how they carry out their functions. He discovered that proteins can be coiled into a spring-like helix, and he came near to proving the structure of DNA. His paper, published with E. J. Corey in 1953, proposed a triple helix for this, rather than the double helix which won Maurice Wilkins, Francis Crick, and James Watson their Nobel Prizes in 1962.

In his home country, Pauling was seen as unpatriotic and too left wing, and in the 1950s he fell foul of the United States government, to such a degree that he was even refused a passport. In 1960 he risked jail for contempt of Congress because he refused to reveal to a subcommittee the names of those who had helped him collect signatures for his anti-nuclear petition.

at the Cardiovascular Research Unit at Edinburgh University in Scotland reported in 1992 that the risk of angina is higher in men with low levels of vitamin C.

Pauling publicized his opinions about vitamin C in the bestsellers *Vitamin C and the Common Cold* and *How to Live Longer and Feel Better*. His advice was to take a total daily dose of 10,000 mg of vitamin C; he claimed it promoted a longer life, improved mental health, and cured infections. He pointed out that our primate ancestors were mainly vegetarian and, by studying what a gorilla ate, he estimated that these primates would consume around 10,000 mg of vitamin C per day. He reasoned that human primates might also benefit from an equivalent amount.

Pauling also contributed to a book by the Scottish surgeon Ewan Cameron, called *Cancer and Vitamin C*, which promoted the idea that ascorbic acid had a major role to play in preventing cancer, although they did not claim it cured a cancer that was already established. In the 1970s, Cameron had treated 100 patients suffering from advanced cancer, at the Vale of Leven Hospital in Loch Lomondside, with 10,000 mg doses of vitamin C, and found that they lived at least twice as long as similar patients in a control group who were not so treated. The two men published these findings in 1976, in the influential US journal *Proceedings of the National Academy of Sciences*. They were challenged by other doctors, however, who found no evidence to support their claims. Indeed, the Cameron study could be faulted, in that the two groups (the test group and the control group) were not screened so as to avoid confounding variables that probably rendered the analysis invalid.

Vitamin C continues to surprise

In 1999, P. Samuel Campbell and colleagues at the University of Alabama in the United States reported that mega-doses of vitamin C could relieve stress, at least in rats. This work supported earlier studies in which elderly women and marathon runners had been tested after taking large amounts of the vitamin, the former showing improved immune functions, the latter showing fewer respiratory infections. In the Campbell study, rats were stressed by being incarcerated in a small wiremesh cage for an hour each day for three weeks. Giving them the rat equivalent of a mega-dose a day of vitamin C protected them against the stress the imprisonment caused. Rats who were given the same treatment but were not given vitamin C suffered weight loss, had lower hormone levels, and had higher levels of antibodies in their blood. Vitamin C may well have similar benefits for stressed humans.

Over the years it has always been assumed that too little vitamin C was harmful, and that any excess was simply not used by the body and disposed of. It now appears that high vitamin C levels may not be such a good thing either, as indicated by research carried out by Ian Blair and co-workers at the Center for Cancer Pharmacology at the University of Pennsylvania in Philadelphia. They reported in *Science* in 2001 that vitamin C might have pro-oxidant tendencies; in other words, it can *increase* the levels of potentially damaging chemicals in the body. These are produced when fatty hydroperoxides react with vitamin C to form unsaturated aldehydes, which can be particularly damaging to DNA. They say these findings may explain why mega-doses of vitamin C do not protect against cancer.

Finally, there are some uses of vitamin C that border on the bizarre. In the 1990s, tights were produced in Japan with microcapsules of the vitamin incorporated in them. These were said to release the vitamin as they rubbed against the legs and were supposed to have a cooling and refreshing effect, and to lead to healthier and more beautiful legs. And if you believe that . . .

The nitrate enigma

In the 1980s there were alarms about nitrates in drinking water, and claims that it caused 'blue-baby' syndrome in the very young, and stomach cancer in the elderly. It now appears that neither claim was justified; indeed, there is evidence that this supposedly dangerous chemical may be part of our body's natural defences.

All living things need protein, and protein is built up from **amino acids** (see Glossary), which have nitrogen as part of their make-up. These acids join together to form peptide links (–NH–CO–), and the products are polypeptides, an alternative name for protein. Nitrogen, therefore, is an essential element for life, and this is reflected in the amount that is present in the body: two kilograms in the average adult.

There is plenty of nitrogen on Earth because it constitutes about 80% of the atmosphere, totalling an incredible 4 trillion tonnes—and all of it useless as a nutrient. The early agrochemists of the nineteenth century assumed that plants had a way of absorbing it directly from the air, and so their early attempts to make general fertilizers failed because they did not include any source of useable nitrogen. When it was realized that plants absorbed nitrogen not through their leaves, but through their roots, farmers began to add nitrogen fertilizers to the soil, and these came either from the guano nitrate deposits of Pacific islands, which had built up from bird droppings over countless generations, or from the mineral deposits of Chilean nitrate.

Only a few microbes and plants have the capacity to 'fix' atmospheric nitrogen, yet, thanks to their efforts over eons, a whole planetary ecology can be sustained. This source of nitrogen will even support continued agriculture if properly managed, but it imposes a maximum on human population density. Crop rotation, combined with compost, animal manure, and sewage, make it possible for a hectare of land to feed ten people—provided they accept a mainly vegetarian diet. On the other hand, the productivity of a hectare of land that is fertilized with 'artificial' nitrogen fertilizer can easily support forty people—and on a varied diet.

It continued to tantalize nineteenth-century chemists that there was a potential resource of this element in the atmosphere, if only nitrogen gas could be turned into something useful like ammonia (NH_3). Attempts to react nitrogen and hydrogen together failed to produce any of this, no matter how they were heated. But if one could get them to react, then the prize was great. Indeed, when it was finally achieved in the twentieth century, it would transform agriculture and enable much more food to be grown on much less land.

The German chemist Fritz Haber struggled for several years attempting to produce ammonia this way until, eventually, he showed it was possible with iron as a catalyst. Then the process engineer Carl Bosch proved that it could be made to work commercially. On 3 July 1909, the BASF company inaugurated the first successful Haber-Bosch chemical plant. Today, there are Haber-Bosch plants around the world producing 150 million tonnes of ammonia a year, most of which goes into making fertilizers, so much so that this input of nitrogen into farmed land now exceeds that of Nature.

Sadly, the first Haber-Bosch plant was not seen as the answer to the world's food supply; rather, it provided Germany with the explosives needed to fight two world wars. The ammonia it produced was used to make nitric acid, and thence explosives. It was only when peace came after World War II in 1945 that the output from such plants could be used to make ammonium nitrate (NH_4NO_3) fertilizer, to the extent that around 2 billion people world-wide now rely on this fertilizer to produce much of the food they eat.

This compound has been one of the main products of the chemical industry for more than fifty years but it is still fraught with danger, especially when ammonium nitrate is stored or carried in bulk—when there is the risk of a massive explosion. The first was at Oppau, Germany, on 21 September 1921, at the first Haber-Bosch plant, when a store of 4000 tonnes exploded, killing 430 work people and local inhabitants. The second devastating blast was at Texas City in the United States on 15 April 1947, when a ship carrying 5000 tonnes blew up, killing 552 people, injuring more than 3000, and destroying a large part of the town. The most recent ammonium nitrate disaster occurred on 21 September 2001 at Toulouse, France, where 300 tonnes exploded. The death toll was a relatively small twenty-nine, with the injured numbering about 650, but the devastation was widely felt. Indeed, half the windows in this city of a million people were blown out.

Plants need nitrate, and there is a natural 'reservoir' of this in the soil, where it is recycled from many sources thanks to microbes and other creatures working on the debris from plants and animals. A little nitrate even arrives with rainwater, which dissolves the nitrogen oxides produced in thunderstorms. Some microbes can abstract nitrogen from the air, and they live in symbiotic relationships with legumes, such as beans and clover. The rhizomes that develop on the plant roots can convert nitrogen gas to ammonia, and they pass some of this to the plant in return for carbohydrate.

Ammonium nitrate fertilizer can treble and quadruple crop yields, but there is a danger of nitrate being leached from the land into rivers because it is highly water soluble. The immediate effect of this is to boost the growth of aquatic plants, making bodies of water unsightly and difficult to navigate, but it may also raise the level of nitrates in drinking waters. Originally, this was thought to be a threat to bottle-fed babies, causing a condition known as methaemoglobinaemia (blue-baby syndrome). In 1972, it was claimed that it caused stomach cancer, and this led to an intensive media campaign, with subsequent legal limits being imposed in the European Union and the United States. (More scientific research has since shown that fertilizer nitrate makes only a minor contribution to soil run-off.)

In the 1950s, several babies in rural communities in the United States were affected by blue-baby syndrome, in which the baby's blood was deficient in oxygen, so losing its healthy red colour. The cause was well-water contaminated with high nitrate levels being used to make up formula feed. Clearly, nitrate posed a threat to human health, and eventually the World Health Organization set a maximum acceptable daily intake of nitrate of 3.65 mg per kilogram body weight.

Nitrate came under the microscope as a possible cause of human cancers of the gut (mouth, oesophagus, stomach, and intestines) when an epidemiological survey in Chile in 1970 linked a high nitrate level in drinking water to these conditions. Subsequent surveys in Europe and North America appeared to confirm these findings, although some investigators found just the opposite, with lower incidences of these cancers among those exposed to most nitrate. Of twenty such epidemiological studies that investigated the links between nitrate and cancer, only two showed a positive correlation while eleven showed no link, and seven even showed a negative correlation, that is, more nitrate meant *fewer* cases of cancer.

Epidemiologists in Sir Richard Doll's group at the John Radcliffe Hospital, Oxford, England came to the conclusion in the 1980s that there was no link between nitrate and cancer. A survey of workers at fertilizer factories, where exposure to nitrate-containing dust is high, showed no increase in cancer rates. It appeared that the earlier epidemiology had been confounded by unrecognized factors, such as other sources of nitrate in the diet, which had not been taken into consideration. Indeed, foods like lettuce, spinach, beetroot, and celery have naturally high levels of nitrate, and vegetables account for about 80% of the dietary intake of nitrate, compared with around 20% from drinking water. Celery has 230 mg of nitrate per 100 grams, spinach has 160 mg, beetroot has 120 mg, and lettuce has 105 mg, and while none of these is likely to be eaten in great quantity, there are some foods, such as potatoes, which are consumed in substantial amounts. It is these, with their 15 mg of nitrate per 100 grams, that probably make up a lot of the nitrate in the average person's diet.

In 1985, it was discovered that human metabolism was itself capable of providing the body with 70 mg of nitrate per day, equivalent to that coming from outside sources. Cells release nitrate in response to infections, and even to strenuous physical exercise such as running and cycling. Moreover, it was always difficult to reconcile the fact that, while the use of nitrate fertilizer increased year by year, the incidence of gut cancers declined year by year. While it is almost impossible to prove that nitrate does not cause cancer, there are now good reasons for thinking it actually protects the human body against disease pathogens. (In the nineteenth century, nitrate had been used in medicines to treat fevers, eventually to be superseded by aspirin.)

The protection that nitrate offers comes as follows: some of the dietary nitrate (NO_3^-) is converted to nitrite (NO_2^-) by specialized bacteria that live on the tongue; this nitrite forms nitric oxide (NO) when it encounters the strong acid conditions of the stomach; the nitric oxide kills harmful bacteria, such as *Salmonella* and *E. coli*, which acid alone may not kill. The earlier theory assumed that production of nitrite was the reason why cancers might be formed because, potentially, nitrite could react with an amine to form an N-nitrosamine, and these chemicals are known to be carcinogenic, as shown by animal tests. It now seems unlikely that this process of N-nitrosamine formation occurs in humans.

Despite the weakness of the evidence that linked nitrate in drinking water to human cancer, restrictions were imposed on the levels of nitrate in rivers, limiting this to 50 parts per million (p.p.m.) in the European Union and 45 p.p.m. in the United States, especially if it was a supply of drinking water. Various treatments were installed at water-purification plants to reduce levels that were higher than this, and restrictions were imposed on the use of nitrate fertilizers in 'sensitive' areas, despite evidence that most of the nitrate in rivers was due to the *natural* action of soil microbes, and not fertilizers. Indeed, 40 kilograms of nitrate is leached per hectare per year this way, whether the land is cultivated or not.

Research into nitrate fertilizers has been carried out at the world-famous Rothamsted Experimental Station, located about 30 kilometres north of London, England, for more than 100 years. Here, researchers have monitored nitrate levels in both undisturbed soils and farmed soils. They were particularly concerned to investigate the uptake of nitrate by crops and soil microbes, and to assess the leakage of nitrate from soils. As a result of their research, this loss can be minimized by changing crop-production practices and linking fertilizer-application levels to crop needs, and by applying it only at certain times of the year. The result is that there need be very little leaching of fertilizer nitrate into rivers and into drinking water.

The nitrate regulations were being based on data that had no sound scientific backing. The European Commission's Scientific Committee for Food delivered its conclusions about nitrate in September 1995, in its document *Opinion on Nitrate and Nitrite*; it came to the conclusion that 'epidemiological studies thus far have failed to provide evidence on a causal association between nitrate exposure and human cancer risk'. Others were equally condemning: 'The directives issued in 1962 by the UN's World Health Organization and Food and Agricultural Organization, and in 1980 by the EU, are now redundant. They need repealing; eventually this will become inevitable', says Dr Jean-Louis L'hirondel of the Regional Hospital Centre in Caen, France, who is a recognized expert on nitrate and health.

In an article published in *Food Science and Technology* in 2000, Tom Addiscott, of IACR-Rothamsted, and Nigel Benjamin of St Bartholomew's Hospital, London, went one step further, arguing that nitrate is actually *beneficial* to the human body, and that it provides an essential defence mechanism against gastroenteritis, which is why it is naturally present in saliva. Indeed, one of the results of removing nitrate from the diet might explain why cases of food poisoning increased so dramatically in the 1980s and 1990s. These authors conclude: 'the EC should arguably stop worrying about limiting our nitrate intake and start making sure that we all get enough'.

Not everyone is convinced that nitrate really is safe, and Peter Weber of the US Center for Health Effects of Environmental Contamination is fighting a rearguard action to keep the limit of 45 p.p.m. set by the United States Environmental Protection Agency. He claims that there is a link between nitrate levels and a range of illnesses such as diabetes, stomach cancer, bladder cancer in elderly women, and non-Hodgkin's lymphoma. The difficulty of measuring dietary intake and metabolic generation of nitrate in individuals makes such conclusions tenuous at best.

There are, however, good grounds for limiting nitrate intake for certain vulnerable groups, such as the elderly. The reason is that, as we get older, our production of hydrochloric acid in the stomach decreases, and this puts us more at risk because less acid means that our body is less effective at dealing with nitrate. The same lack of acid also puts those on a poor diet at risk as well, making the elderly poor even more vulnerable, and there is some epidemiological evidence that the incidence of stomach cancer among the elderly in socially deprived regions is higher where nitrate levels in drinking water are above average, at least in the United Kingdom.

Fertilizing an oilfield

For oil companies, one of the problems is that, at best, only 50% of the oil can be extracted from oil-bearing strata deep within the Earth's crust. Sometimes, as much as 65% of the oil remains firmly stuck to porous rocks such as sandstone. The Norwegians have found a way of overcoming the problem: inject sodium nitrate solution into these underground reservoirs of oil. This encourages the proliferation of nitrate-reducing bacteria, and these in turn loosen the oil from the rocks. By the end of this decade, the Norne oil field, which is under the Norwegian Sea and close to the Arctic Circle, may well be able to yield around 500 million barrels of oil,* more than twice the amount currently thought to be extractable.

* 1 barrel = 159 litres, or 42 US gallons.

CHAPTER THREE

Virility, Sterility, and Viagra

SEX AND PROCREATION may be the driving forces behind much that we do, but the delights of sex, and the fulfilment that comes with raising a family are now overshadowed by deeper worries, especially for men. Are you infertile? Are you having difficulty getting it up? What can you take that will set you and your partner's pulses racing and hopefully rekindle a relationship? And can you really afford to go beyond casual relationships and bear the expense of a lasting commitment? In this chapter we shall look at some molecules that have a role in sex, impotence, infertility, and even in that once-common ritual of courtship and engagement.

Somewhat oddly, the word *chemistry* has quite a different meaning when talking about sex, and people use the term in an approving way in terms of sexual attraction between two people. To fulfil *that* chemistry, however, definitely requires an input from *this* chemistry, and we shall look at five chemicals with a part to play. These are nitric oxide, which spurs a young man into action, Viagra, which can spur an older man into action, amyl nitrite, which can enhance the effects of sex, and selenium, lack of which can explain why a man is firing blanks. The fifth chemical, diamond, is employed outside the body and is used to tell the world that you are in love and are loved. Those readers who are likely to be offended by the first four topics are advised to jump immediately to the fifth topic, on page 105.

Nitric oxide (NO)[20]

This toxic gas, once regarded as an atmospheric pollutant, is the key that turns the valve that allows blood to flood into the penis. It is also the molecule that activates many other bodily processes, some of which may get out of control. But how is this simple free radical formed? And what can it really do?

Nitric oxide regulates bodily activities from our head to our toes. Brain, nose, throat, lungs, stomach, liver, kidneys, genitals, intestines, and blood vessels all need it. It is there when we swallow and when we defecate. It is needed in the fight against viruses, bacteria, and parasites. It is involved in every fleeting thought, every dream, every pain we experience. Every moment of our life, our body generates a constant supply of NO molecules, none of which lasts more than a few seconds. Some medical conditions need the supply of NO to be boosted within minutes, while others need it instantly to be reduced.

The heart needs oxygen in order to generate energy to pump blood, and this comes via the coronary arteries. If these become restricted, generally with fatty deposits, then the patient experiences the chest pains of angina, often as a result of the least exertion on their part. (If the arteries become blocked then the victim suffers a heart attack.) What the arteries need is a quick fix of NO. Some long-standing medicines for the relief of angina are able to supplement the body's natural supply of NO, and thereby boost the flow of blood to the heart by relaxing the muscles of blood vessels. In a man, the same molecule and effect will trigger an erection, and by the same chemical action.

Why the body should use NO seems hard to understand, because this has long been known to be an unpleasant gas which reacts with oxygen to form nitrogen dioxide (NO_2), and this in its turn reacts with water to form nitric acid (HNO_3), a process that in the past has contributed to atmospheric pollution and acid rain. Nitric oxide is a **free radical** (see Glossary), and the body is engaged in an endless battle with such molecules because they can damage cells, cause ageing, and initiate cancer.

[20] Also known as nitrogen monoxide or nitrogen(II) oxide.

Deliberately to generate NO seems counter-productive, but then we are dealing with a rather exceptional molecule.

Nitric oxide was probably first made by Johannes Baptista van Helmont (1579–1644), a Flemish alchemist who led rather a secluded life on his private estate near Brussels. He was more than just a seeker after the Philosopher's Stone and the Elixir of Life, as his papers revealed when his son published them soon after his father's death. Van Helmont knew that there were various gases and, indeed, he was the first to use this word, which he derived from the Greek *khaos* meaning chaos. Although he probably made nitric oxide, it was beyond the techniques available to him to investigate its properties. That was left to the English chemist Joseph Priestley (1733–1804), who devised a method of collecting gases by allowing them to rise up inside an inverted glass vessel filled with water. Joseph Priestley is rightly credited with the discovery of NO in 1772, although his work may well have been based on the earlier observations of the English alchemist John Mayow (1640–79), who wrote about various 'airs', one of which might well have been NO.

There are several chemical reactions that will produce NO but the simplest way, and the one that was used in school laboratories, was to allow concentrated nitric acid to drip on to copper turnings. To begin with, red-brown fumes form in the reaction vessel as the nitric oxide that is produced reacts with the trapped oxygen of the air to form nitrogen dioxide, but this unwanted side-product is dissolved out as the gas bubbles up through the water, and the NO that collects is colourless. A better way to prepare pure NO on a small scale is to react sodium nitrite ($NaNO_2$) with ascorbic acid (vitamin C).

When NO is exposed to the air, it quickly turns red-brown, but this happens only in ordinary air and the reaction is catalysed by water vapour. (Dry oxygen gas does not react with NO.) Chemically, the gas behaves like oxygen in that it will support combustion, although not so well.

Nitric oxide has a certain fascination for chemists in that it is that rare bird, a *stable* free radical. The extra 'odd' electron that it possesses should make it extremely reactive but it isn't, and the gas has been much explored over the years. It can lose this odd electron and become NO^+, known as the nitrosonium ion, and this can even be produced and handled as a sulfate salt. NO^+ has a strong ability to bond to metals, and hundreds of such compounds are known.

NO is produced industrially by the reaction of ammonia (NH_3) and oxygen gas at 900 °C, again a reaction that requires water vapour to be present as well as a catalyst composed of the metals platinum and rhodium . There are two main uses to which the gas is put: to make nitric acid, and thence to manufacture ammonium nitrate (NH_4NO_3) fertilizer; or to make hydroxylamine (NH_2OH), and thence to make nylon. In the former case the nitrogen of the NO is likely to end up in our bodies as protein, while in the latter it is likely to be part of our surroundings, perhaps even clinging to our legs.

In the twentieth century, nitric oxide had an unwanted role as one of the so-called NOx gases emitted from the exhausts of engines. It was partly responsible for the smog that polluted the air of cities in hot climates, and was also a contributing factor to acid rain in temperate climes. Now that both of these problems have been brought under control, the gas has lost its stigma and, indeed, when young people hear of NO, they are more likely to know it for its key role in sex.

NO! Surely not?

So how was NO's role in the body discovered? In fact, it was a missing link in a sequence of events. It was once thought that acetylcholine was the messenger molecule that told the muscles of blood vessels to relax. However, when two American scientists, Robert Furchgott and John Zawadzki, removed the endothelial cells which line the walls of the blood vessels, and with which the acetylcholine interacts, they found that this molecule no longer had the expected effect. Clearly, the acetylcholine was only the first messenger, and it was a second messenger released by the endothelial cells that was activating the muscles. What could this second messenger be? They named it EDRF, short for endothelium-derived relaxing factor, but could find no trace of it.

There the problem rested for the time being, although there were suggestions that EDRF would resemble the drugs that could also relax the muscles of blood vessels, that is, nitroglycerine and amyl nitrite. The common factor these share is an NO_2 grouping of atoms, and clearly this was related to their medical action.

Eventually, it was realized that the second messenger was NO. This was difficult to accept given the nature of this gas, although it was already known that NO could be produced and emitted by bacteria. The idea that

....
NO Nobel Prize

The 1998 Nobel Prize for Medicine was shared by Robert Furchgott, Ferid Murad, and Louis Ignarro for 'their discoveries concerning nitric oxide as a signalling molecule in the cardiovascular system'. It was a subject that was rightly deserving of the prize because they were the first to show that a gas can act this way.

In 1980, Furchgott devised the experiment that proved there was an unknown signalling molecule that relaxed muscles of blood vessels. Murad had found, in 1977, that nitroglycerine worked by releasing NO and that this made muscles relax, and he speculated that it might have a natural role in the body, although he could not prove it. Ignarro carried out the analysis which, in 1986, proved NO was the messenger molecule.
....

higher animals would deliberately produce it, and use it, seemed absurd because NO was an unstable, toxic, free-radical gas, and as unlikely to be generated within the body as winning a Nobel Prize. But it was, and it did —see box.

While the future Nobel Laureates were conducting their research, the pharmaceutical industry also had scientists working on the problem. In the mid-1980s, Salvador Moncada and colleagues at the Wellcome Research Laboratories at Beckenham, England, accepted that NO was the missing messenger, and to study its effects they developed a small-scale version of the equipment used in the car industry for measuring the NO in exhaust gases. Using NO gas directly from a cylinder, they were able to show how this molecule caused muscles to relax. They also found, much to their surprise, that blood vessels could make the NO they needed from the amino acid arginine, of which there is a plentiful supply in the human body. (For more on **amino acids**, see Glossary.) The protein of nuts is particularly rich in arginine and, for example, peanuts contain 11%. Other foods with lots of arginine are peas (9%), rice (9%), meat (7%), eggs (6%), fish (6%), and potatoes (5%). Our blood contains about 14 mg of arginine per litre.

Arginine has a guanidine group as part of its make-up, and it is one of the nitrogens of this part of the amino acid which is plucked off to make NO. The enzyme that can do this is *NO synthase*, and it needs oxygen gas

(O_2) to be present as well, one of whose atoms is added to the nitrogen to form NO, while the other replaces the nitrogen taken from arginine, thereby converting it to citrulline. (The body can recycle this back to arginine.) One of the first pieces of evidence that NO was being produced from this amino acid was the use of arginine labelled with radioactive nitrogen-15, which then ended up in the NO and could be detected by mass spectrometry.

The ability of organisms to produce NO is not a recent evolutionary development—witness the horseshoe crab, with its origins going back 500 million years. This creature evolved a process to make NO from arginine, and it uses NO to stop its blood cells from coagulating.

The Wellcome group was now able to explain how the drugs amyl nitrite and nitroglycerine work.[21] These can stop a painful attack of angina by releasing more NO, which relaxes the constricted vessels that are reducing the supply of blood and oxygen to the heart muscle. Nitroglycerine loses one of its nitro groups, of which it has three, when it encounters an enzyme called *mitochondrial aldehyde dehydrogenase* and so forms the nitrite ion, NO_2^-, and this is easily reduced to NO. Mitochondria can process only small amounts of nitroglycerine this way, however, which explains why the effects of these kinds of vasodilators last only a short while.

Exactly how nitroglycerine works was revealed only in 2002 by work at the Duke University Medical Center in Durham, North Carolina, where a team led by Jonathan Stamler discovered that *mitochondrial aldehyde dehydrogenase* is the enzyme responsible for releasing NO from this molecule. The researchers were also able to explain why repeated doses of nitroglycerine have less and less effect. This is because each enzyme that processes a nitroglycerine molecule is thereby rendered inactive, insofar as it can deal with nitroglycerine, and it is possible to deactivate all the enzymes in the mitochondria.

Nitric oxide was first identified in the brain by John Garthwaite and colleagues at Liverpool University, England, and they showed that this organ makes NO in just the same way that blood vessels do. These findings were confirmed when Solomon Snyder at Johns Hopkins University in the United States cloned the nitric-oxide-producing

---

[21] Amyl nitrite is also known as isoamyl nitrite, and nitroglycerine is also known as glyceryl trinitrate.

enzyme, *NO-synthase*, and found that this was abundant in the brain. In fact, the body has three types of *NO synthase*: one for the arteries, one for the brain, and one for the immune system. That the brain contains more *NO synthase* than any other organ reveals just how important NO is to its functioning.

Nitric oxide might well be the much-searched-for 'retrograde messenger' which is the basis of memory. How does a receptor cell in our brain that has once been stimulated recognize the same stimulation again? The receptor does this by sending a 'message received and understood' signal back to the cell that sent the message, and this in its turn programmes itself to send an even stronger message next time. Researchers have speculated that, because NO is so abundant in the brain, it serves as this messenger, although this has yet to be proved.

Because NO is small, it can diffuse into and out of cells easily, and it is quickly mopped up when its job is done. NO is generated in nerve cells and spreads rapidly, activating all cells in its immediate vicinity. White blood cells, such as macrophages, produce NO in abundance and use it as a war gas against invading micro-organisms. The response may be so massive that it dilates the blood vessels to such an extent that the blood pressure drops and the patient becomes unconscious. That may be the first indication that the person is suffering from so-called septic shock, which can quickly be fatal. Inhibitors that block the NO-forming enzymes can restore the blood pressure within minutes, and these are being used to treat such patients. This condition, of there being too much NO, is rare; what many people experience as they get older is a *shortage* of NO, which manifests itself as heart disease and angina.

Nitric oxide may diffuse in and out of most body tissue quickly and easily, but it cannot pass through blood vessels because it is rapidly destroyed whenever it enters a red blood cell. There it encounters a haemoglobin molecule that is carrying oxygen, a speedy reaction occurs in which the NO is oxidized to nitrite and perhaps even to nitrate.

By itself, nitric oxide may be a relatively stable free radical, but that is in comparison with other free radicals which may exist for only a fraction of a second. Even so, its lifetime in the body is brief, making it almost impossible to observe it in action. Methods were developed for detecting NO at low concentrations, but these were not suitable for studying it *in vivo*. All that changed with the ground-breaking work of Tetsuo Nagano and a group of analytical chemistry colleagues at the Graduate School of Medicine of the University of Tokyo, Japan, which they published in 1998. They had solved the problem of tracking NO by designing compounds that fluoresce when they encounter it.

These compounds are diaminofluorescein dyes, which react with NO and thereby glow with an intense green light, the wavelength and intensity of which can be measured and used to calculate how much NO is present. The reaction with NO is highly specific, and there are no spurious signals formed by the dye molecules reacting with other substances in biological tissue. The test is extremely sensitive, to the extent that it can detect NO at concentrations as low as nanograms per litre, in other words, parts per *trillion*.

Doctor NO

The medical use of NO-forming drugs goes back more that 125 years, during which time they have saved countless lives. The story began with Antoine Jerome Balard (1802–76), who investigated the components responsible for spoiling the odour of *Eau de Vie de Marc*, the brandy distilled from pressed grape pulp after wine has been made. He identified it as amyl nitrite and noted that it caused him to have violent headaches. Others who studied the compound also breathed in its vapour and noted how it caused the heart to beat faster and the face to become flushed, although the effect lasted only a minute or so.

The medical effects of amyl nitrite were investigated by Sir Benjamin Ward Richardson (1828–96), who described them in a lecture to the British Association for the Advancement of Science at its annual meeting in 1864. It remained for another London physician, however, Sir Thomas Lauder Brunton (1844–1916), to deduce that it worked by dilating blood vessels. Sir Thomas tried it on his patients who suffered from angina, and with considerable benefit. He reported this in the leading medical journal, *The Lancet*, in 1867, and the rest, as they say, is history.

Amyl nitrite is a volatile liquid that boils at 98 °C. Those who were at risk of an attack of angina could carry a tiny glass capsule of amyl nitrite, which they would break into a handkerchief and breathe in the vapour to obtain relief. It is mentioned in the Sherlock Holmes story 'The Case of the Resident Patient'.

Meanwhile, in Sweden, at the factory owned by Alfred Nobel, something strange was happening. While Nobel himself complained of throbbing headaches when he came to work, some of his employees who suffered from heart trouble, and who were making and handling nitroglycerine, reported that their chest pains disappeared when they were working. This observation came to the attention of local doctors, and eventually it led to nitroglycerine being prescribed as a treatment for heart disease, to the extent that even Nobel was treated with it when he became ill, despite the fact that he would experience the side effects of headaches. In a letter he wrote at the time, he commented: 'It is ironical that I am now ordered by my physician to *eat* nitroglycerine.' The treatment involved placing a tiny pill of 0.5 mg of nitroglycerine under the tongue. This quickly dissolved and was absorbed into the bloodstream, providing relief within a couple of minutes. (There was no danger of such tablets exploding, because the nitroglycerine was diluted with lactose or some other carbohydrate.)

Exposure to nitroglycerine can cause migraines of the kind experienced by Nobel, and referred to by doctors as nitroglycerine headache. This was only one of a group of side effects experienced by those who handled it, the others being flushing of the face, palpitations, and the passing of copious quantities of urine. The condition was also known as Monday headache, because it generally affected workers on the first day of the working week but passed off as the body adjusted to it.

The effects of nitroglycerine also came to the notice of doctors during World War I when the women who were packing it into shells complained of feeling light-headed and dizzy. When their blood pressures were measured, they were found to be very low, and the reason was that they were absorbing enough through their lungs and skin to have an effect. Trinitrotoluene (TNT) was another explosive which caused similar problems among munitions workers during World War II, and in some cases led to their deaths.

Nowadays, nitroglycerine comes in a spray pack called Coro-Nitro, which is used to spray the tongue immediately an attack of angina begins, or just prior to exertion that is likely to prompt an attack. It can also be applied directly to the skin as a 10 mg or 5 mg patch (Deponit).

All these various forms of medication contain an NO_2 group of atoms, which can be attached to the rest of the molecule through one of the oxygens, when it is referred to as a nitrite, or through the nitrogen atom, when it is called a nitro compound. Normally you would not want to treat people directly with NO because it is a toxic gas, as the great English chemist Sir Humphry Davy (1778–1829) discovered in 1800 when it nearly killed him. Yet, in the right dose it can save lives.

In the 1990s, NO was used in clinical practice as the gas itself; it was given to relieve lung congestion in adults, and even babies, when added in trace amounts (25 p.p.m.) to the oxygen they were breathing. About 20% of babies release a slimy substance called meconium from their intestines into the amniotic fluid, from where it can pass into their lungs and make breathing difficult once they are born. This condition can be relieved by NO gas, resulting in the baby turning from an oxygen-starved blue to a healthy pink within a relatively short time.

Generally, though, the better way to increase NO in the body is to give it indirectly in various ways as nitrate compounds. Those which have been used to treat heart conditions are erythrityl tetranitrate, isosorbide dinitrate, and penta-erythritol tetranitrate, all of which have been prescribed as vasodilators under various trade names; for example, erythrityl tetranitrate was also known as Cardilate. This compound is slow to act, taking fifteen minutes (although its effects last up to three hours), which probably explains why it was less popular than isosorbide dinitrate, which takes only three minutes to work and lasts an hour. This is still widely used and prescribed under numerous names such as Elantan, Isoket, Isordil, and Monit, and is available as tablets, chewable tablets, or mouth sprays. Finally, there is penta-erthythritol tetranitrate, which can last up to six hours but takes twenty minutes to act. This last is now rarely prescribed, although it was once available as Mycardol tablets. All these vasodilators are metabolized by enzymes in the muscle cells or the cells of the blood vessels, and release NO.

All are, like nitroglycerine, highly explosive and are used as such, their power deriving from the several nitrate (NO_3) groups that they contain. When this decomposes it releases energy as it forms nitrogen gas (N_2) and, at the same time, its oxygen atoms combine with carbon and hydrogen atoms in the rest of the molecule, thereby releasing even more energy as they react to form carbon dioxide and water. Hit these compounds hard and the shock wave that passes through the material causes nitrate to collide with nitrate and, within a millisecond, a chemical reaction begins that becomes a chain reaction and an explosion.

The more we have learnt about NO, the more we can put other observations in context. For example, we now discover that NO has been protecting our food for over a hundred years. Meat producers have long used sodium nitrite to inhibit dangerous bacteria from growing on cured ham and in tins of corned beef, although no one knew exactly why it worked so well. Now we do understand: this simple salt acts as a source of NO. Even when we have eaten our corned-beef sandwich, the NO may be there to help it on its way—because this is the molecule that triggers the wave-like contractions of the gut that moves food through our stomach and intestines. Sodium nitrite gives meat a fresh red appearance and it does this by being reduced to NO, which then bonds to the iron of the haemoglobin to form a pink compound.

Macrophages are cells in the blood that seek out foreign particles, such as invading bacteria or mutant cells, and destroy them by injecting a fatal dose of NO. NO binds to proteins called transcription factors that turn genes on and off. The NO attaches itself to sulfur atoms of the amino acids in the protein, which then has to be removed, although this takes time. It is this buying of time which the body can use to gear up its immune system to fight the invading microbes and kill them. Bacteria have their own defences, however, in the form of decoy proteins with sulfur atoms that can mop up the NO, and they even have a transcription factor that, when attacked by NO, responds by switching on a large number of genes to make their own defence proteins.

The binding of NO to the sulfur atoms of proteins is now recognized as one of the more important activation mechanisms that goes on in the body. The research of Jonathan Stamler of Duke University Medical Center, North Carolina, in the United States, has been instrumental in uncovering the way in which NO brings about its effects. We now know that NO targets two types of atom: the sulfurs of proteins, and the metal atoms at the centre of haem molecules. It is by this latter type of interaction that it has its effect on the messenger molecule cGMP[22]—more of this in a minute.

If the body produces a little too much NO, this gives rise to local inflammation. One of the most irritating symptoms of whooping cough is the cough itself, which is caused by the overproduction of NO in the windpipe.

---

[22] Short for cyclic guanosine monophosphate.

No NO, no sex

Men are not the only ones to 'flash' females to reveal their readiness to engage in sex. Fireflies do it all the time, and they too rely on NO. Barry Trimmer and co-workers at Tufts University, Massachusetts, reported this in 2001. The NO is the chemical messenger that triggers the cells that generate the flashes of light, and these researchers found that, when the insects were put into an atmosphere with 70 p.p.m. of NO, they glowed permanently.

In men, the effect of NO is also rather dramatic. Erotic stimuli in the brain send a signal to the nerves of the corpus cavernosum, the spongy muscle in the penis, which then releases nitric oxide. This relaxes the gatekeeper muscle and so lets blood enter the tissues, causing them to swell and produce an erection. The first paper to appear on this role of NO was published by Professor K.-E. Andersson of the Lund University Hospital in Sweden in 1991 in the journal *Acta Physiologica Scandinavica*. (In fact, it had actually been submitted for publication after a similar paper from a group at the University Medical Center, Boston, Massachusetts had been sent to the *Journal of Clinical Investigation* for publication, but this appeared after the Andersson paper.) As you might expect, the work attracted world-wide media attention, but this was as nothing compared to the media blitz that resulted in saturation coverage of our next molecule.

Raising the dead: Viagra

As men become middle aged, their desire for sex can sometimes be frustrated by their inability to maintain or even obtain an erection. And while this may be no big deal when they are old, it can be a very big deal when they are younger. It's then that they turn to Viagra for help.

Why does a man's penis become erect several times during the night, and without his being aware of it? Often he will awaken with an erection. Freud had an answer: he must be having sexy dreams and, if this was not the case, then the dreams he was having must be loaded with sexual symbolism. Had Freud known about enzymes and the effect of nitric oxide, he might well have come to another conclusion: the erect penis was more likely to be causing the dreams and not the other way round.

As men get older, their desire to perform a sexual act can be frustrated by their unenthusiastic member. They are suffering from what is technically called erectile dysfunction (ED), and it is then that they turn to Viagra for help. It used to be said that women could fake an orgasm but men couldn't fake an erection. Now they can; they can literally be cocksure of being able to carry it off, thanks to Viagra.

A British neurophysiologist, Dr Giles Brindley, is famous in the annals of medicine for a lecture he gave on treatments for impotence at a conference in Las Vegas in 1983. He was addressing delegates at the annual meeting of the American Urological Association, telling them how injecting phenoxybenzamine into the penis would produce a remarkable erection. This drug was normally used to reduce abnormally high blood pressure. Perhaps he suspected there were those in the audience who would be sceptical of such a claim, so he had a surprise for them that they would never forget.

Minutes before the lecture began, 57-year-old Brindley had injected his own penis with the drug and it was clearly working, so he stepped from behind the lectern, pulled down the trousers of the jogging outfit he was wearing, and displayed to his audience the wonderful effect it had had. Not only that: he walked around the lecture theatre and invited delegates to feel his penis to prove that it was not being kept rigid with an internal splint (which up to then was one of the ways of keeping an impotent penis rigid enough for penetrative intercourse).

Brindley had discovered the benefits of phenoxybenzamine when a colleague had suggested that drugs which lowered blood pressure might have just the opposite effect on the penis. Nothing daunted, Brindley went home and tested some on himself with remarkable results. In one of his research papers, he described the effects of papverine, which, he noted, produced 'an unrelenting erection lasting *four* hours'. This drug is still used by those who are prepared to inject into the penis, and male actors in porno movies are believed to rely on it.

Brindley's success spurred others into injecting different compounds and, indeed, some were equally successful; one of them was eventually marketed by Pharmacia & Upjohn as the drug Caverject (chemical name

---

NONO, Carl, you may have got it right!

In 1998, Carl Djerassi, the playwright, poet, novelist, and man who invented the contraceptive pill, published a novel with probably the shortest title ever: *NO.* The book is about an Indian scientist, Renu Krishnan, who discovers compounds that can release NO where it is most needed, in the penis, and about the unscrupulous goings-on as she tries to start a new company to produce it. In the book, Djerassi introduces the idea of double potency NONO molecules, in which one NO attaches itself to another NO grouping within an existing molecule, and indeed some of these are now being actively researched and tested as possible drugs. They might be used to promote healing of blood vessels after balloon angioplasty, to relieve pulmonary hypertensions, to inhibit blood clotting, and to preserve donor hearts for transplantation. The difficulty with such drugs is to design them so that they work only in the affected location.

---

alprostadil). It was the first such drug to be approved by the US Food and Drugs Administration for the treatment of ED, in 1995. It worked very well, the only drawback being that it had to be injected into the base of the penis. Not surprisingly, it was almost a treatment of last resort, but then some men were stranded on that lonely island.

Of course, if you were incapable of sticking a needle into your penis, there were other, slightly less painful ways of stinging the flaccid member into action. One of these was to insert a pellet of alprostadil into the tip of the penis and wait for the drug to diffuse along the urethra and into the surrounding tissue, whereupon it, too, would have the desired effect. Such treatments for impotence feature in Carl Djerassi's literary works—see box.

Viagra is the trade name for sildenafil citrate. It was discovered at the Pfizer laboratories in Sandwich, Kent, England in the late 1980s. In a way, the discovery could be said to be accidental, because the team who were making it were seeking a treatment for angina. Simon Campbell and David Roberts began their search for alternative drugs in 1985, and they started looking for a molecule that would block the enzyme *phosphodiesterase,* which has the function of deactivating the messenger molecule cGMP. One of the roles of cGMP is to act as a vasodilator, in other words, to relax the vascular muscles of blood vessels, which it does by causing calcium ions to move out of muscle cells. Nick Terrett joined Campbell and Roberts the following year in the search for suitable molecules.

One compound that was known to work was zaprinast, a molecule with two rings of atoms to which other groups of atoms were attached. The Pfizer team set about changing these groups, with a view to making it more efficient at blocking the enzyme that it was attracted to. Eventually they were satisfied that they had such a molecule, and it was given the code number UK92480 and the name sildenafil.

Laboratory tests showed the compound to be non-toxic, and clinical trials on humans were started in July 1991. Volunteers were given increasing doses of the drug to see whether it had any side effects and, indeed, it had when given in large doses. In some men it caused headaches, indigestion, visual disturbance, and aching muscles, but in almost every man it produced hard erections, which some of those being treated had not experienced for many years. Not surprisingly, this unexpected side effect became the main focus of the tests. Could it produce an erection in much lower doses? It could, given time. Double-blind tests were then performed on a group of volunteers, with the somewhat curious result that about 30% of the men who were given a placebo also reported improved erections.

Of those given a 25 mg dose of sildenafil, 65% reported stronger erections. Those on a 50 mg dose had an 80% success rate, while those on the maximum dose of 100 mg achieved almost 90%. (The other 10 % said the side effects of headache and indigestion were enough to be a turn-off.) These tests were carried out in the United Kingdom, France, and Sweden. By the end of the trials, Pfizer knew they had a winner, because almost all the men on the treatment wanted to continue with it; indeed, many were reluctant to return unused pills.

Clearly, the drug needed to be marketed under a sexy name and the result was Viagra, instantly memorable and, no doubt, its rhyming association with Niagara helped. That gigantic waterfall, famous for its thunderous and copious flow of water, is also a favoured location of honeymoon couples in North America. Within weeks, news of Viagra was splashed across headlines around the world as millions of the pale-

How Viagra works

To produce an erection a man has to be sexually stimulated—by thought, word, or deed—and this releases nitric oxide (NO) from nerve endings in the spongy cells of the penis. This simple molecule then kicks into action an enzyme, *guanylate cyclase,* that produces cGMP, which then relaxes the muscles supplying the penis with blood, and this causes more blood to flow. The penis enlarges, and eventually becomes so engorged with blood that it stands erect.

Meanwhile, another enzyme called *phosphodiesterase* is there to remove cGMP, albeit at a rate that cannot cope with the rush of NO and cGMP. As a man gets older the body's production of these chemicals is insufficient to counteract the neutralizing effect of the *phospho- diesterase* enzymes, and the result is impotence in the form of only partial erections or unsustainable erections. Viagra corrects these dysfunctions by blocking the *phosphodiesterase* enzyme.

But why should Viagra affect only the penis, when cGMP production and removal are used in many other parts of the body? The answer lies with the *phosphodiesterase,* of which there are several varieties, the one in the penis being *phosphodiesterase-5.* Viagra deactivates this while leaving other phosphodiesterases, such as the one in heart muscle (*phosphodiesterase-3*), unaffected. Viagra is just the right molecular size to fit the active centre of *phosphodiesterase-5* and block it. Until the enzyme can free itself of its unwanted burden, it is inactive against its real target molecule, cGMP, and so the level of this builds up in the penis and continues to be high while the man remains sexually aroused.

The side effects of Viagra, such as headaches and feeling dizzy, are due to the dilation of blood vessels in the brain. Another side effect of Viagra is temporarily to give some men blue-tinted vision, and it does this because the cone cells responsible for colour vision also rely on *phosphodiesterase-5.*

blue, lozenge-shaped pills poured off the Pfizer production lines. The blue Viagra pill also contains other ingredients such as cellulose, calcium phosphate, titanium dioxide, lactose (to bulk it out and make it disintegrate quickly in the stomach), and a blue dye so that it will be a distinctive and recognizable colour. It is then stamped with the letters VGR25, VGR50, or VGR100 to show the dose it will deliver.

Because its effect is at the molecular level, Viagra will normally work whatever the underlying reason for a man's ED, be it depression, stress, the result of another disease such as diabetes, or because of prostate surgery.

The famous 1940s' investigation into the sexual life of Americans, carried out by the Kinsey Institute, had shown that 15% of men over fifty were impotent. Things have clearly gone downhill since then and, by the 1990s, research papers were claiming that as many as 40% of 40-year-old men experience some degree of ED. Those figures came from a small study on 303 men carried out in 1994, and they were considered to be suffering from ED even if they reported this as 'minimal'. Nevertheless, it was publicity that Pfizer was happy to exploit and, in any case, was almost impossible to verify. The prestigious United States National Institute of Health estimates that as many as 30 million American men suffer some form of ED, and that only one man in five seeks treatment for his affliction. The publicity surrounding Viagra emphasized that ED was a much more common complaint than formerly recognized.

Viagra is not an aphrodisiac, and it works only in response to sexual stimulation. In theory, if you take Viagra and are not so stimulated it will not work, although it is hard to imagine a man who had taken Viagra who would not then be thinking about sex.

Viagra was launched in the United States in 1998, where it had been tested on two groups of 532 and 329 men, to be followed by much larger trials of several thousand. In these trials two men died, one aged sixty-six who was a heavy smoker, and the other fifty-three, who was apparently in good health. These deaths were not unexpected considering they were dealing with a group of middle-aged men.

During the first week of its launch more than 35,000 prescriptions were dispensed; within three weeks they totalled 300,000, and by the end of the year had exceeded 5 million. Indeed, it became the fastest-selling drug in pharmaceutical history. It also had an equally arousing effect on Pfizer's shares, which shot up from $45 to $115 within two weeks. By the summer of 1998, hundreds of thousands of US males had used it, and sixty-nine of them had died as a result. Of these, however, forty-six were men with heart conditions, which having sex might well have exacerbated.

A Viagra pill is swallowed about an hour before sexual activity is undertaken and will boost penile potency for three to five hours. An effective erection leading to ejaculation can be produced several times during this period. Viagra can be used as a recreational drug, and is so used by some, but it can be risky, especially if taken in conjunction with poppers, the street name for volatile nitrites. The combination of these two drugs can cause such a dilation of the arteries that cardiovascular collapse is a real threat.

Not all the side effects of Viagra are negative—as far as rhinoceroses and harp seals are concerned they have been very positive. The African rhino was facing extinction, because rhino horn was an important component of Chinese medicine and was supposed to be particularly effective in curing impotence. Now Chinese men take Viagra. (Of course, rhino horn would work on the 30% of men who responded to the placebo in the Viagra trials.) Rhino horn is a single-strand protein called keratin with no possible medicinal benefits, or at least none that you could not get by ingesting other forms of keratin, such as those from ground-up animal hooves or even finger nails. In any case, the Chinese government banned the import and use of rhino horn in 1993; with a bit of luck rhinos will not now join the list of extinct mammals.

Although they were not put at quite the same risk of extinction as rhinoceroses, male harp seals were also being killed for part of their body: the penis. These, too, were supposed to be a cure for impotence, and by the time Viagra came on the scene they were fetching $100 per organ. After Viagra appeared, the price dropped dramatically to $15, and so did demand.

Because Viagra was such a money spinner, it was hardly surprising that other drug companies followed suit with their own versions, even claiming to come up with better and safer products. Schering-Plough launched an anti-impotence drug, Vasomax; Eli Lilly and Icos came out with Cialis; and Glaxo-SmithKline sells Levitra. This last drug underwent trials on a group of 805 men, 75% of whom reported an erection sufficient to effect penetration after taking a 10 mg pill. Levitra is said to target the *phosphodiesterase-5* enzyme more efficiently, and it has fewer side effects: there is no effect on the eyes, for example. And while it may take a little longer to work its magic, Levitra will last for up to twelve hours compared with Viagra's four hours. It appears to be particularly effective for men who have undergone prostate surgery, which often leaves erectile dysfunction in its wake.

Pfizer challenged these alternative drugs in the courts, and though the challenges were ultimately unsuccessful, it was able to protect its patent on Viagra up to 2013. What weakened the Pfizer case for exclusive rights to drugs that interfere with *phosphodiesterase* was that several researchers had published work in this area in 1992 and 1993, and this convinced the High Court judge that Pfizer had been merely putting into practice the recommendations of others, even though the original Pfizer patent was dated 1991.

Love potions and poppers

Although no pharmaceutical company manufactures aphrodisiacs as such, there are some chemicals, nominally prescribed for other conditions, which appear to have this as a side effect. And there are some that are reputed to produce the ultimate orgasm.

According to the dictionary, an aphrodisiac is something that arouses sexual desire. It isn't a stimulant to engorge the penis, but something that renews a desire for sexual intercourse when this desire is lacking for some reason. The male fantasy of an aphrodisiac is something that will generate a lust for sex in both himself and/or his partner. Humans have hankered after them for thousands of years—witness the recipes for love potions that have been preserved on Egyptian papyri from the time of the Middle Kingdom, around 2000 BC.

There are chemicals that appear to act as aphrodisiacs, such as yohim-bine, bromocriptine, and deprenyl. The first of these comes from the inner bark of a West African tree, *Corynanthe yohimbe*, and has been used to initiate and prolong sexual arousal in men. The molecule is an indole alkaloid, and because it is a herbal remedy it can be procured from suppliers of such products. It works in a different way to nitric oxide in that it appears to increase the flow of blood into the penis while decreasing the flow of blood out of that organ.

Bromocriptine is one of the ergot alkaloids and, as its name implies, it contains a bromine atom. It has therapeutic properties and can be used in the treatment of Parkinson's disease. (It is also reputed to be useful in other treatments, such as for obesity.) It acts to suppress the release of the hormone prolactin from the pituitary gland which, as its name implies, triggers the production of breast milk and continues to maintain it as long as a woman is breast feeding. Prolactin is present in men and women, and the more prolactin we have in our body the less we feel like engaging in sex. Bromocriptine stimulates the level of the feel-good chemical, dopamine, in the brain, and it is this that accounts for its ability to cause sexual arousal. By suppressing prolactin, bromocriptine has thereby something like an aphrodisiac effect.

Deprenyl is also used to treat Parkinson's disease and is a relatively simple molecule compared with yohimbine or bromocriptine. It has similarities to amphetamine and to phenylethylamine, the chemical that is supposed to give chocolate its addictive power. Deprenyl has one unusual feature, a carbon-to-carbon triple bond, but whether this has any connection to its reputed aphrodisiac effects is not known.

Amyl nitrite is no longer prescribed for the treatment of angina, but is nevertheless still available and still widely used to enhance orgasm. Indeed, for gay men it is second only to cannabis as a way of enhancing sexual performance. No longer sold in ampoules, nitrites now come in screw-top bottles which are still referred to as 'poppers' (the name originally came from the 'pop' that accompanied the snapping of an amyl nitrite ampoule as the vapour escaped).

Nitrites can be sold in the United States, but only as solvents for cleaning things such as leather and video heads, because their possession as poppers is illegal. In most other countries, they can be bought quite legally in sex shops. Nor are poppers these days likely to be amyl nitrite, but rather butyl nitrite or isobutyl nitrite, which are even more volatile. Butyl nitrite boils at 78 °C and isobutyl nitrite at 67 °C, whereas amyl nitrite boils at 98 °C, and it is the lower boiling point that explains why isobutyl nitrite is regarded as the most potent of all.

Their effect is short lived, but if the vapour is breathed in during sex it allows more blood to flow into the penis, making it bigger and harder, and the result is referred to as the ultimate male orgasm. In the United States, the FDA declared these nitrites safe, but they were banned in 1991 because it was thought by the government that their use by gay men might be a factor in the spread of AIDS. Poppers enhance the pleasure of anal sex by relaxing the muscle of the anal sphincter. (In some respects this action may even be beneficial, in that there is less likely to be damage caused and so less likelihood of HIV transmission between partners.)

All these love potions and pleasure enhancers ultimately rely on the generation of NO in the body, even if this enhancement comes by stimulating the level of feel-good molecules in the brain (for more on these, see Chapter 5).

An element of sex: selenium

Young men today are supposed to be less good at producing sperm than their fathers, and suggested causes range from sedentary lifestyles to endocrine disrupters, but there is one factor necessary for male fertility that tends to be overlooked, and that is selenium.

A man who lacks selenium produces lazy sperm and less of them, and while he may not be aware of a lack of this nutrient in his diet, he will soon become aware of the lacklustre performance of his seed when he tries to father children. The link between selenium and male fertility was demonstrated by a Scottish researcher, Alan MacPherson, in 1993; he reported on a double-blind trial in which selenium supplements were given to a group of men, and their sperm counts monitored. The men who were given a placebo continued to have low sperm counts, whereas the men who took the selenium pills soon doubled their output of viable sperm.

Obviously there are many reasons why a man cannot father children, but selenium may have become a factor because, in some parts of the world, the amount of this essential element in the average diet is well below that recommended. Men no longer prefer to eat selenium-rich cuts of meat and, in Europe, they no longer get bread made from American wheat, which has lots of selenium.

Until 1975, selenium was not seen as an essential element for the human body. It was useful for things like light-sensitive detectors and photocopiers, but it was regarded as rather toxic and, indeed, men exposed to selenium in their occupations would soon find their friends and partners avoiding them. This element can be absorbed through the skin and is then expelled through the sweat glands, leading to repulsive body odour, and through the lungs, producing the most foul-smelling breath that it is possible to exhale. While such men might be producing Olympic swimming sperm, their chances of these finding their target were virtually non-existent.

Then, in 1975, our attitude to selenium had to change: it was an essential element after all. Yogesh Awasthi, based at Galveston in Texas, discovered that it was part of the antioxidant enzyme *glutathione peroxidase*, which destroys peroxides before they can form dangerous free radicals. In 1991, Professor Dietrich Behne, at the Hahn-Meitner Institute in Berlin found selenium in another enzyme *deiodinase*, which promotes hormone production in the thyroid gland and other tissues.

We now know that a molecule of *glutathione peroxidase* contains four selenium atoms and that, as a consequence, every cell of our body contains more than a million atoms of this element. The average American male has in his body around 14 mg[23] of selenium, most of which is to be found in his bones, although those parts of his body with the highest levels are the hair, the kidneys, and the testicles. The average British man has only around half this amount.

Even though selenium is essential, we can have too much, and the recommended *maximum* daily intake is 0.45 mg. Above this, we risk selenium poisoning, with its repulsive symptoms which are caused by volatile methyl selenium molecules that the body produces in an effort to rid itself of an unwanted excess. But there's the rub: too little selenium and a man may be infertile; too much selenium and he becomes repulsive to be near. (A single dose of five grams would make him seriously ill.)

The ideal daily intake of selenium is 0.2 mg, which is the same as 200 *micro*grams, this being the unit of weight that nutritionists prefer to use when talking about it. This may seem tiny but, nevertheless, even this

---

[23] 14 mg of selenium represents a total of 10^{20} atoms, which is 100 million trillion atoms.

minute amount can be lacking from the diet.[24] How much we take in can vary widely and, while we have a pool of selenium that the body can draw on temporarily, in the long run this needs to be supplemented with at least 50 micrograms of selenium a day. If our average daily intake is much less than this, then eventually the body store will be depleted and the selenium will be conserved for use in only the most vital parts of our metabolism, of which producing happy sperm is not one.

Most people get a daily dose of selenium from breakfast cereals and bread, but there are richer sources of the element, such as nuts; the Brazil nut is particularly well endowed, with 20 micrograms of selenium per nut, although this can be very variable. Eating three of these a day should easily provide the necessary minimum of selenium. Indeed, a single selenium-rich nut per day may well be all you need to transform your diet from deficient to sufficient. Another nut with a relatively high selenium content is the cashew, and a 100-gram bag of these can contain 65 micrograms.

Although the level of selenium in the sea is very low (0.2 parts per billion (p.p.b.)), some fish are able to concentrate it and, of these, the tuna, cod, and salmon have quite a lot. Offal, and especially liver and kidney, are also good sources, although these cuts of meat are no longer popular, and fewer men today eat liver and bacon, or steak and kidney pie, both of which would have boosted their fathers' gonads.

What is most likely to provide selenium in the US diet is wholemeal bread, four slices of which will yield 60 micrograms; indeed, all wheat products and breakfast cereals contain some selenium. Mushrooms have a little; a portion of fried mushrooms (say 50 grams) would have 5 micrograms but, were these to be the edible mushroom *Albatrellus pescaprae*, which is popular in Italy, then such a portion would provide more than 1000 micrograms. These are safe to eat because much of the selenium in mushrooms is not easily digested.

For those who want to boost their selenium intake with supplements, and seek only natural ones, then brewer's yeast is the best, provided this has been grown on a selenium-rich culture medium. For those who simply want to pop in a pill in the morning, there are many on the market, and they generally contain selenium methionine (see below), although

---

[24] A *lifetime's* supply is only around five grams, which is about a small teaspoonful.

some contain the simple salt sodium selenite ($NaSeO_3$), which is the chemical form given to animals in areas where the soil lacks selenium. Sodium selenite is a white crystalline material that is soluble in water and is easily absorbed by the body, but there are components in the diet that will reduce its absorption, such as a high level of fat and protein.

Selenium was discovered in 1817 by Jöns Jacob Berzelius (1779–1849) at Stockholm, where he was the Professor of Chemistry and Medicine at the Medical School. In the summer of that year, he was asked to investigate a red-brown sediment which collected at the bottom of the chambers in which sulfuric acid was being made at Gripsholm, a works in which he had invested. His first thought was that the sediment contained tellurium, because of the strong smell of radishes it gave off when he heated it. (Tellurium had been discovered twenty years earlier at a mine in Romania, where it was found as the mineral, gold telluride.)

Further research convinced Berzelius that it was not tellurium, but rather it was a new element, albeit very similar in its chemical properties. He named it selenium, from the Greek, *selene*, meaning 'Moon', to complement the name tellurium, the name of which was derived from the Latin, *tellus*, meaning 'Earth'. Berzelius noted that the new element was very like sulfur in that it would vaporize on heating and re-deposit as 'flowers' on a cool surface. It was different, however, in that flowers of sulfur are bright yellow whereas those of selenium are red. This is the non-metallic form of the element. We now know that sulfur, selenium, and tellurium are in the same group of the periodic table, hence their chemical similarities. Berzelius paid a price for his curiosity and research: he began to smell so much that one day his housekeeper accused him of eating too much raw garlic because his breath smelt so bad. He also put his life at risk when he was overcome by breathing in hydrogen selenide gas (H_2Se), which is dangerously toxic.

Nowadays, selenium is extracted from the slime that settles at the bottoms of tanks in which copper is refined electrolytically. The main producing countries are Canada, the United States, Bolivia, and Russia, and the main consumers are the United States, Europe, and Japan. Production runs at 1500 tonnes a year, of which 10% is recycled from industrial waste or reclaimed from old photocopiers. Selenium is a metalloid element and it can exist in two forms: a metallic form and a non-metal form. The metallic form of selenium has the curious property of

> ### Contact lenses
>
> Contact lenses coated with selenium metal may one day be on the market, and their advantage would be in allowing them to be worn for several weeks at a time. The selenium would serve two functions: it protects the eye from bacterial infection by producing powerful oxidizing agents, and it prevents microbes from attaching themselves to the lenses.

increasing its electrical conductivity by a thousandfold when light falls on its surface, and this accounts for most of its usage, and why it is needed for photoelectric cells, light meters, and solar panels. These electronic uses require high-purity (99.99%) selenium, and they account for a third of production. The second largest use is in glass, either to decolourize it or to produce the bronze architectural glass that screens out the sun's rays. The third most important use is as sodium selenite for animal feeds and food supplements, and these utilize about 250 tonnes a year.

Shortages and surpluses of selenium

Selenium is a rare element here on Earth, but rarity does not exclude some soils from having too much. The amount in the atmosphere is only about one nanogram (a billionth of a gram) per cubic metre, and comes as methyl selenide and dimethyl selenide, which are released from soils, lakes, and sewage by anaerobic bacteria. Just how important this is came to be realized in 1989, when the oil company Chevron stopped discharging selenium-polluted waste water into San Francisco Bay. Chevron pumped it instead to a 35-hectare wetland area that had been specially created to deal with the toxic waste; it was thought that most of the selenium would be extracted, ending up in the sediment. Somewhat surprisingly, when the company scientists monitored the sediment they found only half the selenium that they expected. The rest had simply been digested by microbes, turned into gases, and carried away on the wind.

It has been estimated that 20,000 tonnes of selenium vent into the atmosphere each year, of which two-thirds are due to human activity, such as coal burning and metal smelting. A lot of this may even be beneficial, at least the part which is washed down by rain on to selenium-deficient land. Soil that has been well fertilized over the years contains about 400 mg of selenium per tonne, because the element used to be present in the widely used phosphate fertilizer known as superphosphate; this was manufactured from sulfuric acid (H_2SO_4), which contained traces of selenic acid (H_2SeO_4). In some soils the level of selenium has been deliberately increased by adding selenium salts.

Some parts of the world have soils with naturally high levels of selenium, to the extent that the plants growing on them can be toxic to grazing animals, leading to a condition known as blind staggers. The famous Venetian traveller Marco Polo (1254–1324) observed this when he travelled east along the Silk Road to China, and reported seeing animals that behaved as if they were drunk. The Great Plains of the United States are also rich in selenium, and the cowboys of the Wild West were also familiar with the blind staggers; they attributed this to their herds feeding on milk-vetch, which they called locoweed from the Spanish word, *loco*, meaning 'insane'. They were right to do so because milk-vetch (*Astragalus*) has an ability to absorb large amounts of selenium, to the extent that it comprises more than 1% of the plant's dry weight. In 1934, a biochemist, Orville Beath, finally proved that that blind staggers was due to too much selenium.

Nor is it only livestock that are affected by selenium; wildlife can also suffer. In 1983, the Kesterson Reservoir in the San Jaoquin Valley, California, became so polluted with selenium from agricultural drainage that wildfowl chicks were being born deformed, and some adult birds were even dying because of it. The drainage canal that fed into the reservoir was cut off and reclamation undertaken.

Too much selenium may be bad, but too little can be even worse. The country with the largest area of selenium-depleted soil is China, especially the Keshan and Linxian regions.[25] The lack of selenium means that there are only minute traces of this essential element in the diet of those

---

[25] Virus mutations often appear in China, and one theory links this to low selenium levels in the population. This reduces immunity to disease and, according to the theory, allows virus populations to grow much larger than usual, making it more likely that a virulent strain will emerge. In a selenium-deficient individual the normally benign coxsackie virus can become virulent and cause damage to the heart, and an infection by the influenza virus can lead to more damage to lung cells.

who live there, and those most at risk were the children of Keshan. They were prone to a type of congestive heart failure that became known as Keshan disease, which causes an abnormal swelling of the heart and kills half of those afflicted. The cause was suspected to be the lack of selenium in their diet and, in 1974, a large-scale trial was undertaken involving 20,000 children; half were given selenium-containing tablets and half a placebo. Of those on the placebo, 106 developed Keshan disease and 53 died, while of those taking the selenium supplement only 17 contracted Keshan disease and 1 died.

Another large-scale experiment, this time in Linxian, indicated that lack of selenium was a factor in the high incidence of stomach cancer. A five-year project was undertaken with 30,000 middle-aged people, who were given different combinations of vitamins A, B_2, C, and E, along with zinc or selenium. The study showed a remarkable drop in cancer cases among the group taking vitamin E and selenium together. Investigation of whether lack of selenium was also affecting male fertility was not part of the study, but this was hardly an issue at a time when couples were limited to only one child by law. The apparent lack of selenium in the Chinese diet appears not to have acted as a brake on population numbers in any case.

In the United Kingdom, farmers have been giving their animals selenium supplements for years to keep them healthy. Land in Britain lacks selenium partly because it was washed from the soil by the melting ice at the end of the last ice age, 10,000 years ago, and partly because of the notoriously high rainfall since then.

When, in 1984, the Finnish authorities decided to put selenium in all their fertilizers, it was because the average diet in that country was delivering only 25 micrograms per person per day, barely above the 20 microgram level needed to prevent Keshan disease. What also convinced them that there was something wrong was the incidence of white muscle disease in cattle, and the high incidence of heart disease and cancer among the general population. As a result of the government's decision, the level of selenium in the farmland of Finland is such that the dietary intake of the population is now more than 90 micrograms per person per day.

The dilemma of selenium, as with many other elements, is that, if a person has either too little or too much of it in their diet, his or her health will suffer. The daily intake of selenium varies between 6 and 200 micrograms, depending on the type of foods that are eaten. The average American takes in as much as 90 to 120 micrograms per day, which is more than adequate to prevent selenium deficiency, while in the UK the intake is only a third of this, 30 to 40 micrograms per day, much less than the recommended intake of 75 micrograms for men and 60 micrograms for women.

Selenium in foods can come in the form of selenocysteine, which occurs in things such as broccoli and garlic, or as selenomethionine, which is how it occurs in meat and grain. Selenocysteine has one carbon atom fewer than selenomethionine and does not get incorporated into protein, whereas selenomethionine can be incorporated into protein and is less available. Selenocysteine is supposed by some to be superior in terms of cancer prevention.

Selenium appears to function as an antidote for metal toxins, especially cadmium, mercury, arsenic, and thallium, and it is known technically as an antagonist for these metals. Tuna, which accumulates higher-than-expected levels of mercury, is thought to be safe to eat because this fish is partly protected by the selenium that it also extracts from its food.[26] In the 1970s, there was a food scare in the United States when analysis of canned tuna revealed higher-then-expected levels of mercury, which were assumed to be caused by environmental pollution. The selenium content of tuna, and its protective role, were not then understood. Overnight, canned tuna disappeared from supermarket shelves and millions of cans were destroyed. The analysis of a sample of tuna from the nineteenth century that had been preserved in a museum display case, however, showed the same level of mercury. The protective role of selenium appears to be true for other marine mammals, such as seals, and may also be the way that the men who work in mercury mines can adjust to this dangerous metal, provided they don't lack selenium.

Health benefits of selenium

A lack of selenium in the diet is said to be linked to other conditions besides Keshan disease and male infertility: high blood pressure, cancer, and arthritis have also been mentioned as possible risks. The evidence is

[26] Even so, for safety the amount eaten should not exceed one tuna steak or two cans of tuna per week.

largely epidemiological, and selenium is likely to be only a secondary factor, but in some cases the evidence seems compelling. If the underlying cause is due to free-radical damage, then a lack of the selenium-containing enzyme that destroys these would be a disadvantage. Trials have shown that supplementing the diet of those exhibiting the early signs of Alzheimer's disease slowed down its rate of progress, at least insofar as there was less of a decline in cognitive ability.

In one test in the United States involving 1300 elderly people over four years, there was a 30% decrease in new lung, bowel, and prostate cancers among those who took a selenium supplement. It appears that selenium may be needed to activate a gene that makes a natural chemical which suppresses tumours.

These days, selenium (in combination with the antioxidant vitamins A, C, and E) is widely taken as a health supplement. Selenium therapy was first popularized by Alan Lewis in his book *Selenium: the Essential Trace Element You Might Not be Getting Enough of*, which was published in 1982; in writing this, he was well ahead of his time. Alerted to the problem by this and other publications, a whole selenium supplement 'industry' has grown up to cater for those who want to benefit from taking selenium. There are even those who advocate taking 'organic' selenium as opposed to 'inorganic' selenium, and they aim to boost their intake of those amino acids that have incorporated this element into their chemical make-up, and especially selenomethionine. Some farmers produce selenium-rich eggs by giving hens selenium-rich feeds, and indeed such eggs can contain more than 20 micrograms of the element.

One of the leading experts on the role of selenium and health is British scientist Margaret Rayman, who is based at the Centre of Nutrition and Food Safety at Surrey University. She wrote an editorial in the February 1997 issue of the *British Medical Journal* claiming that the problems most associated with low selenium levels are infertility, cancer, and heart disease. Rayman pointed out that in the past twenty-five years the intake of selenium by the average Briton has fallen from an almost adequate 60 micrograms a day to an inadequate 35 micrograms a day. The United Kingdom's Cancer Research Campaign agreed to fund a pilot project in 1999, at a cost of £210,000 and involving 500 volunteers, which Rayman organized. It was hoped that this would be followed by a large-scale project in which more than 10,000 individuals would take part, in a series of double-blind tests. Scientists at Cancer Research UK agreed to collaborate but, in 2002, when the full project was submitted, Rayman found to her dismay that the main funding body, the UK's Medical Research Council, lacked the £6 million needed to carry it out.

> ## Selenium gets a boost
>
> The action of selenium can be boosted by other molecules, and one that appears to have a particularly enhancing effect is sulforaphane, a naturally occurring compound that is present in broccoli, cabbage, cauliflower, and Brussels sprouts. While both selenium and sulforaphane individually protect against cancer, together they exert a much more powerful anti-cancer effect. Sulforaphane contains an isothiocyanate group of atoms and, as such, is an antioxidant. Collaborative work between the United Kingdom's Institute for Food Research, China's University of Science and Technology, and Scotland's Edinburgh Royal Infirmary have shown that this molecule and selenium have a synergic effect; in other words, each boosts the biological activity of the other and, together, they boost the body's detoxifying enzymes.

Healthy sperm needs selenium

No man likes to think he's firing blanks, even if his gun is in working order and, thanks to Viagra, the latter should not pose a problem. While chemical research found the answer to ED, it has yet to come up with a tonic for the testicles, one that will ensure only healthy sperm are produced and ejected. Nevertheless, that does not mean nothing can be done about it.

According to some researchers, such as Alan MacPherson, mentioned above, the declining sperm count may be explained by a decrease in selenium in the diet. There is good reason to believe this, because sperm needs selenium to develop properly. Sperm require extra selenium in the nucleus, in the form of *glutathione peroxidase*, and in the sheath surrounding the mitochondria. Selenium is abundant particularly in the mid-section of sperm, which joins the head to the tail, where it appears to have some structural function. Men with low selenium levels tend to produce a lot of sperm with inferior motoring and navigational skills, which might well be linked to a lack of selenium at this vital point. This role for the element was discovered in the sperm of rats as long ago as 1972, by D. B. Brown and R. F. Burk, and published in the *Journal of Nutrition*. There is every reason to believe that a human sperm needs selenium just as much as does that of a rat. Other tests on the sperm of mice, boars, and bulls came up with similar findings. Brown and Burk injected male rats with the radioactive isotope of selenium, Se-75, which has a half-life of seventeen weeks, and were able to show that this appeared in the part of the sperm between the head and the tail.

In another experiment, seven young cockerels were raised on a low-selenium diet, but with a very high level of vitamin E to see if this could compensate. It did to a limited extent, but only two of the seven cockerels produced mature testicles; the other five were infertile. In another experiment, boars were fed on a diet of selenium-deficient corn and soybean meal that had been specially grown in an area of China where the soil lacked selenium. The food of this diet had a level of selenium of only ten parts per billion. Half the boars developed undersized testicles, producing less than half the usual sperm count and with three-quarters of those having malformed heads and tails.

Men, of course, are not boars, nor even rats, bulls, or cockerels (although they may take on the characteristics of some of these animals), so perhaps none of this research applies to humans. Almost certainly it does, though, because they all produce sperm and it is unlikely that Nature would change the blueprint just for humans.

Clearly, there can be lots of factors that explain why a man cannot produce a baby, and lack of selenium is likely to be the cause of only a fraction of the cases, but it may be as high as one in ten. This was shown in double-blind tests in which one group of infertile men was given a selenium supplement, and then their success rate in fathering a baby was compared with another group that was given a placebo: the selenium group showed an increase of 11%.

For the remaining 89%, the cause may be one of many factors, including lifestyle. Cornell University in the United States has an Institute for Reproductive Medicine, and it issues general guidelines that infertile men should follow. These are to avoid hot baths, hot tubs, and saunas; to limit the intake of coffee to two cups a day; not to smoke or use recreational drugs such as marijuana; not to drink more than 60 grams of alcohol per week (equivalent to three pints of beer or six glasses of wine); to eat lots of fresh fruit and vegetables; and to exercise regularly. They also point out that vitamins C and E may help improve fertility, the object being to keep antioxidant levels high because free radicals can destroy the membrane that surrounds a sperm, and it has been shown that infertile men generally have higher levels of free radicals in the semen. Zinc can also play a part, because the level of this in semen is high, and zinc-containing enzymes are also important in fertility. A daily dietary intake of 200 micrograms of selenium is also recommended.

OK, so you are fit and healthy, eating a balanced diet, with no reason to think you are infertile, and you've now fallen in love with the woman you want to marry and raise a family with. How do you tell her, and the world, that this is the real thing? What better way than to give her a diamond ring. And if you think that this is still nothing more than the romantic kind of chemistry, then think again.

Diamonds are a girl's best friend

. . . so sang Marilyn Monroe, who was supposed to have been the most desirable woman of her day, and who even bedded the US President, John Kennedy. This precious stone is still regarded as the perfect symbol of a man's undying love for a woman.

The 17th of July 1986 was a momentous day at the Premier diamond mine in South Africa. There they extracted one of the greatest diamonds ever found and, although in the rough state it was irregular in shape, it tipped the scales at 599 carats (almost 120 grams). It was a near flawless crystal or, in the language of the diamond trade, a diamond of the first water. A team of diamond cutters spent the next three years designing and building the equipment to fashion the stone, and it took nine months of patient work to extract from the rough diamond the largest possible regular gemstone. Eventually, however, they were able to display a perfect gem of 273 carats, with 247 facets and an unmatched brilliance and fire. It was named the Centenary diamond because its existence was announced to the world in 1988, exactly 100 years after the

De Beers Company began mining in the area. What the diamond is worth, or even who owns it, is anyone's guess; these are closely guarded secrets of De Beers, but it seems unlikely that it will ever grace the person of a Hollywood star, however desirable she may be.

Diamonds have come to symbolize everlasting love, summed up by the memorable advertising phrase 'A diamond is forever'. Indeed, diamond is a material of extremes and, for that reason, it is as attractive to chemists as to lovers. Diamond is pure carbon; it is the hardest naturally occurring material known, and cold to the touch. Less well known is that each diamond is a single molecule, and that it won't last forever. Diamonds can not only be manufactured, they can also be used to coat all kinds of things—and they can be any colour. Some of the most famous diamonds are coloured, such as the blue Hope diamond, which resides in the Smithsonian Institution in Washington DC.

Many of the largest stones have histories. The Koh-i-noor diamond, for example, was discovered in India in the thirteenth century and is probably the world's most famous gem. It was acquired by the British after the second Sikh war in 1849, when it had already been cut down to 186 carats. Even then it did not please Queen Victoria because it didn't sparkle enough, so it had to be recut, which reduced it to its present 109 carats. Smaller it may be, but is still priceless and now resides in the Queen Mother's crown in the Tower of London. Other great diamonds are individually named, and among those rare stones are the Tiffany, the Cullinan, the Dresden Green, the Taylor–Burton, the Millennium Star, the Shah Jahan, the Sancy, and the Regent. The Regent resides in the Louvre in Paris, and was bought by Philippe d'Orleans, Regent of France in 1717, for £135,000 from Thomas Pitt, a merchant and president of Fort Madras. Pitt had acquired it in a rather shady deal for £20,000. To equate these sums to modern prices you need to multiply by 100.

It may seem odd that diamond, strong though it be, is not the most stable form of carbon. It seems all the odder when you consider its chemical bonding. Every carbon atom in diamond is bonded to four other carbon atoms in a giant three-dimensional lattice so that, in effect, the complete gem is connected from one facet to the other by direct carbon-to-carbon bonds, thereby making the whole stone a single molecule. These bonds are strong, which is why the whole structure is so rigid and diamond so hard, and why diamonds do not burn like other forms of carbon—

unless, of course, they are heated to very high temperatures, when they will slowly react with oxygen and disappear as carbon dioxide.

Yet, diamond is intrinsically unstable with respect to the other common form of carbon, graphite, although the conversion rate is negligibly slow, and to an extent that diamonds will never be a wasting asset however long they are kept. In graphite, every carbon is bonded directly to only three other carbons in a planar, honeycomb-like array, and these are stacked in layers one above the other. Although there are fewer carbon-to-carbon bonds, they are stronger than those in diamond. There is also a fourth interaction between the layers of graphite, although this is so weak that it is easily separated, which is why graphite is easily cleaved and is as soft as diamond is hard.

Diamonds have enriched the imagery of language with their attributes of beauty, hardness, coldness, purity, fire, wealth, and brilliance. They also offer a sort of financial security in times of crisis because they can easily be hidden and transported.

A popular term for diamonds is 'ice', and you might think this is because they resemble it in being a colourless solid, but that is not the reason. Diamonds are called ice because, when a large diamond is pressed against the tip of the tongue, it always feels cold. The reason for this lies with its high thermal conductivity, the stone quickly draining heat to itself. This property explains why diamond is being researched as a 'heat sink' for microchips.

Although diamonds are described in terms that link them to water, such as 'first water' and 'ice', they are as unlike water as it is possible to be. In fact, you cannot wet a diamond with water because it is effectively an organic macromolecule, although this means you can coat it with oil, and that diamonds will actually stick to fat, a fact known to the ancients.

The Greeks of the sixth century BC knew of diamonds, which they called *adamas* (unconquerable), and they knew that they came from the East, although exactly where was a mystery. In fact, they came from India. Rumour had it that you could get diamonds from deep shafts in the earth by dropping down a chunk of meat fat tied to string, hoping that when you pulled it up again there would be diamonds stuck to it. There is even a story that Alexander the Great (356–323 BC) succeeded in obtaining a few diamonds from a deep gully by throwing down pieces of fatty meat to which some of the gems adhered. When birds swooped down and picked up the pieces of meat, they brought diamonds out with them. This ancient fable has a ring of truth, and this method of winning diamonds is not very different from one that was introduced in the diamond-mining industry in the 1890s. This employed a greasy table, over which a slurry of ground-up diamond ore was flushed. Only the diamonds stuck to the grease.

As mentioned above diamonds can be any colour: brown (some shades euphemistically being referred to as 'champagne' or 'cognac', depending on the depth of colour), violet, orange, blue, green, pink (rare), blue (even rarer), and red (rarest of all). Violet diamonds get their colour from hydrogen atoms within the lattice. All diamonds have a hint of colour, but only about one in a thousand is intensely coloured, and they command the highest prices. The colour of blue diamonds is caused by traces of boron, while green diamonds are produced by the Earth's natural radioactivity, and this process can be replicated by exposing colourless stones to the radiation of a nuclear reactor, when they take on a spectrum of hues.

The early chemists of the eighteenth century were unable to analyse diamonds, and puzzled over what they were made of, although they knew that they could be destroyed by extreme heat. In 1694, Guiseppe Averani and Cipriano Targioni of Florence, Italy, showed that when they focused sunlight onto a diamond using a large magnifying glass the gem eventually disappeared. But it was not until 1796 that an English chemist, Smithson Tennant, proved that diamond was entirely a form of carbon, and that when it was burned it formed only carbon dioxide. This being so, it was deduced that graphite ought to be convertible to diamond. Attracted by the lure of an apparently easy route to wealth, many tried—and failed.

We now know that this can be done only by using very high temperatures and intense pressures that were quite beyond the technologies of the nineteenth and early twentieth centuries. The Swedish company ASEA made the first synthetic diamonds in 1953, using temperatures of 3000 °C and pressures of 90,000 atmospheres. Early claims to have made diamonds must have been false; yet one eminent French chemist, Henri Moissan, even exhibited a small diamond, 0.7 mm in size, which he said he had made, but we can be fairly certain it was slipped into his apparatus by a laboratory assistant.

Diamonds are mined in Botswana, Russia, South Africa, Canada, Namibia, Angola, and Australia, and some of these stones, about 5% of production, are fit only for industrial uses, which now consume more than 1 billion carats (250 tonnes) per year. Most manufactured diamonds are made in Russia and China. It is possible to produce gem-quality diamonds, but few are because they are costly to make, although demand may yet be stimulated—see box.

A surface film of diamond can be applied to materials by a technique known as chemical vapour deposition. In this process, methane gas (CH_4) is decomposed to its atoms in a microwave furnace, and these then crystallize out on to the surface to be protected (such as a razor blade) in a structure like that of diamond. The resulting film of carbon is invisible to the naked eye but it gives the surface the same qualities of strength and hardness as those of a real diamond.

If diamonds are forever, then even part of your loved one can last forever

That's the claim of Chicago company LifeGem, which specializes in converting carbon from cadavers into diamonds. When a loved one dies and is cremated, it is possible during part of the cremation process for the supply of oxygen to the furnace to be reduced, thereby causing incomplete combustion and the formation of soot (carbon). This can then be collected and converted to diamond.

The soot is first heated to convert it to graphite, and then sent to a laboratory in Germany where it can be subjected to the high temperatures and pressures needed to convert it to a gemstone. This is then sent back to the United States to be returned to the bereaved family members, and perhaps even converted to jewellery. The first person to have their remains so transformed was a 27-year-old woman who died of Hodgkin's disease in September 2002: from her remains six half-carat diamonds were produced. (The costs of LifeGem's memorial stones work out at about $4000 per quarter-carat diamond.) To make it provable that the stone really did come from the ashes of the loved one minute traces of other elements are added to the graphite before it is processed into the diamond.

Whatever industrial uses are found for diamonds, and however easy it is to make them, a large well-cut stone still symbolizes true love. But why should a woman regard the giving of a diamond as a pledge of her man's undying love? Part of the answer is high fashion, and part is popular culture. Diamond rings were worn as betrothal rings in the fifteenth century, the stone being a fitting symbol of one's eternal love. Clearly, however, they were designed only to grace the fingers of the very rich. The modern type of engagement ring can be traced back to the early twentieth century, when E. F. Cushing wrote in his book, *Culture and Good Manners*, that a solitaire diamond engagement ring was the epitome of good taste; by then, the cost of such a ring was within the reach of most love-sick men. Moreover, it indicated the extent of his wealth and the degree of his commitment. The giving and receiving of such a ring are still regarded as announcing to the world that a man and woman are deeply in love.

The popular lure of diamonds can be traced back to a more cynical manipulation of public taste by those in the diamond industry, who were always willing to oblige Hollywood studios with lavish diamond jewellery for any producer or director who wished to feature them. So it was that *Gentlemen Prefer Blondes*, *The Pink Panther*, and even Hitchcock's *To Catch a Thief*, sprinkled diamonds across the screen. No doubt, though, the best publicity of all came as the title of the 1971 James Bond movie: *Diamonds are Forever*.

CHAPTER FOUR

Germs Warfare

Kitchen. Once a room sparsely furnished in which food was prepared with skill, devotion, and hard work. Now a room in which food is rarely prepared and even then with little skill, devotion, or hard work. To compensate, the room is filled with expensive and largely non-active 'appliances' that are worshipped as objects in their own right.

THAT IS A DEFINITION from the amusing *Dictionary of Dangerous Words* compiled by Digby Anderson. He might well have added 'but a lack of hygiene' to his description of the traditional kitchen, and 'an obsessive cleanliness' to the modern one. Yet, somewhat perversely, the incidence of food poisoning is still a real risk even today. Most of us expect to experience the occasional bout of vomiting and diarrhoea caused by it, although it must be said that these are often likely to be caused by eating food we have not prepared ourselves.

There are bacteria other than ones which cause food poisoning that threaten our health; in addition, we may expect to be laid low with one or two viral infections per year, and may perhaps even find ourselves attacked by fungi. So what can we do about it? The obvious answer is to kill the germs that menace us and, while there are various ways of doing this, the most effective and rapid way is to poison them. We have an armoury of chemicals with which to do it, and this is what this chapter is all about. Before we look at specific agents, however, it might help to take a look at the pathogens (disease-causing agents) that cause disease, and perhaps even ask ourselves if it is always the best policy to try to live in an antiseptically clean environment.

Microbes invade our body via the water we drink, the food we eat, and the air we breathe, and they attach themselves to us every time we touch something or somebody. Because most of them are not pathogens, their presence goes unrecognized. Indeed, some are even beneficial because they help us to digest our food, and some even make vitamins that we can absorb from our intestines. Today, we are rarely threatened by the germs that cause serious diseases, but it was not always so.

About 2.5 million people lived in London in 1850, and in that year there were 48,557 recorded deaths, of which 26,325 were due to microbial infections; not that this was realized at the time to be the real cause of those deaths. Compare this situation with that which pertains today. In England and Wales, with a population of around 45 million, there are only 3500 deaths from such causes, although there are still about 35,000 cases of infectious diseases every year. In 1850 things were truly awful, with infant mortality due to infections accounting for the fact that 15 % of babies never made it to their first birthday. Today, that figure is down to 0.5%.

Such changes came about through discoveries in the sciences of chemistry, medicine, and biology, and everyone has benefited, not just the rich and powerful. Progress was painfully slow, although by the late eighteenth century chemists had discovered substances that could prevent microbial infection. Because microbes were still decades away from being discovered, however, there was no reason to use them as such. But sometimes they were used, albeit for the wrong reasons. For instance, it was believed in the eighteenth century that epidemics and disease were caused by breathing an invisible vapour, which was referred to as 'miasma' (from the Greek word for obnoxious waste), and that that came from cesspits, dungheaps, and decaying rubbish. Such smells could be removed using oxidizing agents, although there was still a tendency simply to disguise the malodour with even stronger, pleasanter scents. Masking the smell with a herbal or floral fragrance was thought to be effective, and in times of plague this had been done by holding a pomander of sweet-smelling herbs to the nose.

Of course the theory was wrong, but if oxidizing agents removed the cause of a foul smell then the chances are that they also removed a source of infection. It was only when the miasma from excrement and rotting wastes was banished to sewers that real progress was made. That, and a clean water supply, ensured that the population was no longer exposed to quite so many harmful pathogens, and even those that were around could be minimized by simple hygiene.

Killing germs

We want to eliminate the pathogens that invade our everyday environment and make us ill, but do we need to kill all other microbes as well? Perhaps we can be too clean for our own good, and living in a completely germ-free environment may not be the best option for a healthy life.

Germs, in the form of bacteria, viruses, and fungi, can invade our body, and either colonize some of our organs, creating a condition that requires medical treatment, or release toxins to which our bodies react violently, which is what causes food poisoning. In the former type of invasion, our immune system has special white blood cells that rush to our defence with the ability to produce a range of chemicals that will destroy the invaders. While a healthy immune system deals with a few such invaders every day, at certain times the invaders come *en masse*, overwhelm our defences, and have the upper hand until we produce enough antibodies to defeat them—meanwhile we are ill, sometimes very ill.

Treatment then consists of taking the right kind of antimicrobial medicine, such as antibiotics for bacterial infections, anti-fungals for fungal infections, and even anti-virals, although these pathogens are much harder to destroy once they have invaded the body. While such products are often the result of years of careful research by chemists in the pharmaceutical industries, they are too problematical for the general public to use them without medical supervision (for example, it is no good taking penicillin to cure a viral infection). What we can do, however, is to zap the microbes before they have a chance to zap us.

When it comes to reducing exposure to harmful germs, there are two methods we can use: cleaning and/or disinfecting. Cleaning removes visible dirt and lots of other unwanted microbes along with it, while disinfecting also removes *invisible* dirt and particularly bacteria, especially those pathogens that cause disease.

In which locations are pathogens most likely to lurk? In the home we tend to think they will be in the lavatory area or around the kitchen sink. Outside the home we are most likely to be concerned about the germs in swimming pools, pubs and restaurants, public lavatories, and hospitals. We may also suspect other locations even though we rarely, if ever, visit them, such as mortuaries, abattoirs, laundries, milking parlours, food-processing plants, air-conditioning units, and cooling towers. All can be ideal breeding grounds for harmful germs.

The enemy comes in many shapes and sizes. Viruses are tiny, around a tenth of a micron (a ten-millionth of a metre) across, and are visible only when viewed with the aid of an electron microscope. Bacteria are larger, though still invisible to the naked eye, and consist of single cells between 1 and 20 microns in size. Some have thick cell walls that can be stained by a process named after Hans Gram,[27] while others cannot be so stained and these have much thinner walls. Those which can be stained are called Gram-positive, those which cannot are Gram-negative. Fungi range from single-celled organisms, such as yeasts, with cells about five microns in size, through filamentous moulds, to larger more complex structures such as mushrooms. Some yeasts which normally live within our gut can invade other parts of the body, causing diseases.

The Gram-positive bacteria that we are most likely to suffer from are: *Clostridium perfingens, Staphylococcus,* and especially *Listeria,* all of which cause food poisoning; *Enterococcus,* which is abundant in faeces; *Propionobacterium acnes,* which causes acne; *Staphylococcus aureus,* which makes wounds turn septic or which causes boils; *Streptococcus pyogenes,* which gives us a sore throat; and *Streptococcus mutans,* the ever-present oral bacterium in dental plaque that can lead to tooth decay.

The Gram-negative bacteria include: *Escherichia coli,* which lives in our intestines; *Klebsiella,* which is responsible for hospital infections; *Cam-pylobacter,* which also causes food poisoning, as do *Salmonella enteritidis* and *Salmonella typhimurium;* and *Shigella sonnei,* which results in dysentery.

Fungi are also a threat in the form of *Candida albicans,* which causes oral and vaginal thrush; *Epidremophyton* spp. and *Trichophyton* spp., which manifest themselves as athlete's foot; the relatively innocuous *Malassezia furfur* of dandruff; and the more serious *Microsporum* spp., which lead to ringworm.

Viruses are perhaps the most likely to attack us, and especially those

[27] Hans Gram (1853–1938) was the Danish bacteriologist who first described the technique, in 1884.

such as rhinoviruses, which cause the symptoms we refer to as a cold, or the more serious influenza virus. Another common viral infection is herpes simplex, which leads to cold sores around the mouth and genitals.

Germs can be killed in non-chemical ways. For example, heat will do it, which is why it is always best to cook thoroughly food that has been stored for some time, and to cook it at temperatures that will destroy microbes, that is, above 70 °C. Ultraviolet radiation will destroy microbes; so will gamma radiation and, indeed, this is probably the best way to preserve food for long periods. Sadly, it is less used than it might have been because of intensive campaigns against it in the last century. Its opponents played on the ignorance of the general public, many of whom were misled into thinking that irradiating food would actually make it radioactive.

While such methods of killing pathogens are really for specialist situations, the general public can fight microbes with common chemicals. The weapons in our armoury are quite varied: oxidizing agents, such as chlorine bleach, hydrogen peroxide, and ozone; chloro-phenolics, such as triclosan and PCMX; and the less well-known quaternary ammonium salts, such as cetrimide. These chemicals work in various ways: oxidizing agents react with the molecular structures of microbes; chloro-phenolics can diffuse through cell membranes and block microbial metabolism in various ways; and quaternary ammonium salts attack the cell membrane.

The oxidizing agents are perhaps the most effective, and these attack not only the cell membrane but also proteins and any molecule with either an N–H or S–H bond—and there are many such bonds which are the keys to the functioning of enzymes. The chloro-phenols also attack cell walls and enzymes, and, if enough of these molecules enter a bacterial cell, they can even cause the cellular fluid to congeal into a useless mass. Quaternary ammonium salts are attracted to the cell wall of a bacterium and then puncture holes in it, so that it leaks and can no longer function.

Anti-microbial chemicals have obvious benefits, but they also have drawbacks, as the following table shows. This also includes alcohol (ethanol), another common chemical with anti-microbial properties, which is used in mouthwashes—although not at the concentration given in the table. Alcohol works by dissolving the membrane of a bacterium

and denaturing its proteins, and the best solution to use is one of 70%
alcohol content. Alcohol is also active at 50%, albeit less so, but this
means that in an emergency it is possible to use neat vodka or gin as an
antiseptic.

Antiseptic chemicals, the benefits and drawbacks

Chemical	Benefits	Drawbacks
Hypochlorite bleach	Kills all microbes	Strong 'chlorine' smell
	Has a long shelf life	Dangerous if mixed with other acid-based cleaners
		Manufacture opposed by environmental groups
Hydrogen peroxide	Kills most microbes	Short shelf life*
	Environmentally benign	Destroyed by common enzymes
Ozone	Kills all microbes	Does not last long
		Cannot be adapted to general use
Organo-phenols	Kill bacteria	Manufacture opposed by
	Long lasting	environmental groups
Quats	Kill bacteria	Not very effective against viruses
	Long lasting	and fungi
Alcohol (70% solution)	Kills most microbes	Flammable
	Completely stable	Toxic

* Hydrogen peroxide can be stabilized by storing it in opaque plastic bottles, and by adding citric acid and
thymol; the non-transparent container filters out UV rays, which can decompose it. Nor does the surface of a
plastic bottle decompose it, whereas that of glass bottle does. The citric acid and thymol bind to metal atoms
that can catalyse its decomposition.

When pathogenic bacteria breed in sufficient numbers on our food,
and we eat it, then we are in for a nasty few days. In the United Kingdom,
there are around 85,000 cases of food poisoning that require medical
attention every year. This represents about 15 persons per 1000 of the
population. Of course, most people who go down with food poisoning
have a relatively mild attack, which manifests itself as vomiting and diarrhoea; there may be twenty times as many who experience these symptoms and who do not go to see their doctor. A survey carried out in the
Netherlands in the early 1990s suggests that this may still be an under-

estimate. Although there are only a few hundred registered cases of gastro-enteritis a year in that country, which is renowned for its cleanliness, a survey suggested that there may, in fact, be more than 2 million such incidents, suggesting that only one person in a thousand feels the need to seek medical treatment when afflicted with vomiting and diarrhoea.

Food poisoning can require hospitalization, and it may even be fatal, especially when it affects the very young and the elderly. The bacterium *E. coli O157* comes into this category because the toxin it produces, vero cytotoxin, can cause bleeding in the intestines and even result in damage to the kidneys. Moreover, the bacteria can have a long incubation period of up to two weeks, and as few as 100 organisms may be sufficient to cause the disease. We rarely come into contact with this bacterium, which resides in the intestines of cattle, but if raw beef is contaminated from this source and then not properly cooked it can get into the human food chain. Inadequately cooked minced meat, hamburgers, and unpasteurized milk are the main sources of infection, although children have caught the disease just by stoking animals or picking up sheep droppings. Thankfully, cases are rare, and in the United Kingdom there are fewer than 1000 per year. The most likely causes of general food poisoning are *Salmonella* and *Campylobacter* bacteria, which can contaminate poultry, raw eggs, and dairy products, and it often arises through exposure to the pathogen in a victim's own home or in a restaurant.

One major source of germs is a damp kitchen cloth, which might have as many as a billion microbes living on it. As such, it is an excellent agent for spreading these to every surface it wipes. Of course, most of these microbes are harmless and it is only when such a cloth has been contaminated, say after wiping up chicken blood, that it should be either discarded or soaked in bleach.

Log kills

Bacteria will multiply rapidly provided they have a source of nutrition, moisture, and warmth. On clean, dry surfaces they die, which is why simple cleaning may often be all that is necessary to control them. Viruses multiply only when they invade a living cell, and that happens only if they gain access to the human body. Shake hands with an infected person, or touch a surface they have touched, and then transfer a few viruses to your own mouth or eyes and you risk infecting yourself.

Moreover, viruses can survive for days on a surface, especially if it is dirty, but on a clean surface they will eventually disintegrate. Fungi can be single-celled organisms but most are more complex; they tend to flourish in damp conditions, and are less of a threat.

Bacteria are counted in their millions and billions. Killing them with a chemical agent can be very effective, and the numbers destroyed are so large that a logarithmic scale is needed to describe the slaughter. Thus, we talk of a log4 kill, or even better, of a log6 kill. The former is equivalent to killing 99.99% of a bacterial population, the latter the equivalent of killing 99.9999%. Of course, the few that remain can again replicate rapidly, given a source of nutrients, and become millions within a day.

Log kill sounds odd until you realize that these logs are the logarithms of large numbers. Suppose we have a million bacteria (that is, 10^6 in numerical terms) and kill 90% of them, then we are left with 100,000, which is 10^5, and this is referred to as a log1 kill, which is the numerical difference in the powers of ten of going from 6 to 5, and is a slightly different way of saying 'divide by 10'. (A log1 kill is about what you achieve when you wash your hands with soap and water.) Killing 90 % of the remaining 100,000 will reduce the population to 10,000, which is 10^4, and this is referred to a log2 kill, which is another way of saying we have destroyed 99% of the original million. (Really thorough hand washing can achieve this level of elimination.) Kill 90% of these and we are now down to 1000 (10^3), having achieved a 99.9% kill, or a log3 kill. A further 90% removal gets us to 100 (10^2), and means that 99.99% have been destroyed, that is, a log4 kill. To eliminate 90% of the remainder sees the number of surviving microbes down to 10, in total a log5 kill, and removing 9 of these takes us to a single bacterium and a log6 kill. The net result is that 999,999 of the original million are now dead, and the kill rate has been 99.9999%. There can be log7 and log8 kills if we start with a *billion* or more bacteria, but one assumes that a log6 kill is adequate for most disinfection purposes.

A normal healthy person will need to be infected by several thousand viruses, or more than 100,000 bacteria, to be made ill by them. In susceptible groups of the population, such as children, the sick, and the elderly, however, as few as a thousand microbes may cause illness. Given the right conditions, micro-organisms can multiply at an alarming rate: one bacterium can become a million bacteria in less than 24 hours.

Can we be too clean for our own good?

It goes without saying that modern methods of hygiene save lives, and this is nowhere better reflected than in the dramatic fall in infant deaths over the past two centuries. But could we now be too hygienic? During the span of time it took *Homo sapiens* to evolve, our ancestors relied on their immune systems to protect them, and babies born with deficient immune systems would have been weeded out as soon as they stopped breast feeding and were cut off from its rich supply of protective chemicals and antibodies.

One school of thought now believes that we are in danger of protecting babies and young children to such an extent that their immune systems do not properly develop. That, then, is the reason their immune systems overreact to non-threatening materials such as the enzymes from pets, plants, and house dust mites, and this is why there is so much asthma and allergy about. Contact with such materials is necessary if we are to develop and maintain a network of specialized immune cells, and there are findings that appear to back this up. For example, younger siblings are less likely to develop asthma, hay fever, or eczema than their elder brothers and sisters, and this may be because they are exposed to all the dirt that these older children bring into the home. Recent opinion, however, has veered towards identifying lifestyle factors, such as dietary changes, as more likely to be the cause of allergic diseases.

When the immune system is functioning properly, special cells send out cytokines to instruct infected cells to kill the bacteria or viruses that are harming them. Another set of cells is there to protect against parasites attacking the wall of our gut, and these cells do this by releasing antibodies and activating mast cells to repel the invader. The mast cells release massive amounts of histamine, which provokes a flood of mucus to dislodge the microbes. This response can be triggered by a false identification and then we may become allergic to whatever that is, be it animal fur, pollen, or peanuts.

Normally all we need to do is clean our home rather than disinfect it. There are times, however, when we feel vulnerable to infection, and it is then that we need the extra protection of something stronger. That is when we turn to products that can rid us of disease pathogens. The following table lists the main household disinfectants and their effects on the four types of microbe that plague us.

Disinfectants and their effects on the types of microbe

Disinfectant	Effect on				Mode of action
	Bacteria	Viruses	Fungi	Spores	
Hypochlorite	Kills	Kills	Kills	Kills	Oxidizes vital molecules
Chloro-phenols	Kill	Kill most	Kill	Resistant	Denature key proteins
Quats	Kill some	Kill a few	Kill	Resistant	Puncture membranes
Hydrogen peroxide	Kills	Kills many	Kills a few	Can kill*	Attacks key molecules, e.g. DNA

* At high concentrations.

Hypochlorite

So-called chlorine bleach contains not chlorine but hypochlorite, and this has the power to kill all germs stone dead. Despite its benefits there have been campaigns to ban it.

Chlorine is a toxic, greenish-yellow gas that will kill in minutes. It affects eyes and lungs at a concentration of only 3 p.p.m. in air; at 50 p.p.m. it is dangerous to breathe even for a short time; and at 500 p.p.m. it can be fatal in less than ten minutes.

Chlorine is much less dangerous when it is dissolved in water, with which it reacts to form a mixture of hydrochloric (HCl) and hypochlorous acid (HOCl). It is much more soluble in sodium hydroxide (also known as caustic soda, formula NaOH) solution, reacting to form sodium hypochlorite (NaOCl), which has powerful germicidal properties, and is commonly referred to as household bleach or, somewhat misleadingly, as chlorine bleach—misleading because it contains no dissolved chlorine gas as such.

Chlorine has an interesting history. It was first produced in 1774 by the 32-year-old, German-born chemist Karl Wilhelm Scheele (1742–86) at Uppsala, Sweden, and he made it by heating hydrochloric acid with the mineral manganese dioxide (MnO_2). A dense, greenish-yellow gas was evolved, which he recorded as having a choking smell, and which dissolved in water to give an acidic solution that bleached litmus paper and decolourized the leaves and flowers of plants. Scheele called the gas dephlogisticated muriatic acid,[28] and it was known by this curious name for more than thirty years, until the 29-year-old English chemist Humphry Davy investigated it in 1807 and concluded that it was not an acid but an element.

Meanwhile, others had already found uses for it. In 1786, James Watt of Birmingham, England, had demonstrated the bleaching of cotton with chlorine dissolved in water, but its use was not a practical possibility. It offered an alternative way of bleaching to that normally used, however, in which textiles were laid in strips in fields to be bleached by sunlight, a process that took several weeks. In 1787, the French chemist Claude Louis Berthollet (1748–1822) had absorbed chlorine gas into caustic potash (potassium hydroxide solution, KOH), to form a solution of potassium hypochlorite (KOCl). He called it 'Eau de Javelle' after a small village near Paris, which was a centre for textile bleaching. Soon a factory was set up to produce the new bleaching agent, and it also came to the attention of those engaged in paper manufacturing.

In 1799, chlorine was absorbed into a slurry of slaked lime (calcium hydroxide, $Ca(OH)_2$) to form solid calcium hypochlorite, and this became known as bleaching powder. The benefit of this was that it could be transported easily to where it was needed, when it would be dissolved in water to give a solution of hypochlorite that could be used for bleaching or cleaning.

In 1820, a French chemist, Antoine-Germain Labarraque (1777–1850) started producing 'Eau de Labarraque' by dissolving chlorine gas in a solution of NaOH to form sodium hypochlorite. This was much cheaper to produce than 'Eau de Javel', and this we can still buy today as chlorine bleach, sold under a variety of names that have become so popular that they are part of the language. Thus, in the United Kingdom people speak of Domestos and in the United States of Chlorox, and a bottle of such a product is to be found in many kitchens, bathrooms, and lavatories.

[28] Hydrochloric acid was then known as muriatic acid, a name that persisted in some industries well into the twentieth century. Although we think of HCl as a dangerous chemical, it is produced in the human stomach as part of our digestive system.

Hypochlorites were originally regarded as transportable forms of chlorine itself because they will immediately release this gas when they are acidified with hydrochloric acid; and in so doing they yield up about a third of their weight in the form of gas. Today, it is more economic to transport chlorine as a liquid under pressure in special tankers.

Millions of tonnes of hypochlorite are still manufactured every year, the amount being difficult to determine because often they are made where they are needed, such as in effluent treatment. Household bleach accounts for about a quarter of the total production of chlorine and, despite opposition to its manufacture from environmentalists, it remains a major industry. An unexpected use of bleach is in oil production, where it is used to sterilize seawater that is pumped down into oil deposits to force up more oil. If the water is not sterile, it allows slime moulds to grow and these clog up the oil-bearing strata, making extraction difficult. (In Chapter 3 we also saw how encouraging the right bacteria might even be used to extract more oil.)

In some countries, bleach is extremely popular. In Spain the average person gets through more than 12 one-litre bottles per year, while in other countries it is extremely unpopular; in Germany, for example, it would take a person a whole lifetime to use this amount of bleach. In the United Kingdom, the average person buys two such bottles per year. Bleach cannot be stored for too long because it slowly decomposes to inactive agents such as sodium chlorate ($NaClO_3$) or oxygen gas (O_2), a process that is aided by sunlight. This is why bleach is always sold in opaque containers.

While some people find the odour of bleach unpleasant, many regard its fresh smell as reassuring and, in the early nineteenth century, it offered a chance of banishing 'miasma'. In 1827, Thomas Alcock published his *Essay on the Uses of Chlorites of Oxide of Sodium and Lime*, in which he specifically recommended bleach and bleaching powder for deodorizing particularly smelly environments, such as hospitals, stables, lavatories, and so on. Two years earlier, in 1825, Labarraque himself had experimented with bleach, washing wounds with it, and recommending it be used wherever illness was present, but the idea did not catch on.

Then, in May 1847, a Hungarian surgeon, Ignaz Semmelweis (1818–65), used it to disinfect the hands of medical students in the Vienna Lying-in Hospital (Maternity Hospital) where he worked. He had noticed that the incidence of puerperal fever, which begins as a vaginal infection and eventually causes blood poisoning and death, was killing more than 30% of mothers in the ward where medical students were trained. In an adjacent ward, which was staffed by midwifery students, the death rate was much lower. The cause lay with the medical students, who would often go to the ward immediately after dissecting a corpse, and Semmelweis reasoned that they were bringing some agent of infection with them. Thanks to his insistence that everyone must wash their hands in bleach, the death rate from puerperal fever soon fell to only 1%. (Today, the disease affects less than 0.0001% of women who give birth.)

But the hypochlorite success story was only beginning, and it was to prove instrumental in ending water-borne diseases later that same century. In 1881, the pioneering German bacteriologist Robert Koch (1843–1910) showed how effective bleach was in killing such pathogens, and in 1892 in Hamburg it was used to disinfect drinking water as a means of ending an outbreak of cholera. Five years later, it was added to the water mains of Maidstone, England, to bring an outbreak of enteric fever (typhoid) under control. In 1905, a much more serious outbreak of the disease in Lincoln was likewise ended with the chlorination of its water supply.

It was clear that water-borne diseases were eradicated by this method, so that more and more chlorination plants were installed to purify the drinking water of towns and cities, and they are now standard throughout the world. Indeed, such has been its benefit to humans that, in 1998, *Life* magazine ranked it among the 100 greatest achievements of the last millennium. Hypochlorite has a major advantage that should not be overlooked: it is cheap enough for even the poorest households to purchase, and for the poorest communities to install.

Hypochlorite has been used in other ways, too. Dakin's Solution, consisting of a 0.5% solution of sodium hypochlorite, was used extensively in World War I to disinfect wounds and prevent gangrene, while a slightly stronger product, Milton fluid, which is a 1% solution, became popular as a general antiseptic. For many years this was used to sterilize feeding bottles for babies, with the instruction that, after soaking in Milton, the bottle should simply be drained of surplus liquid and not rinsed before being used again. (The amount of hypochlorite a baby would thereby have ingested would have been tiny.)

The solid sodium dichloroisocyanurate (NaDCC) is now produced in preference to bleaching powder, and it, too, dissolves in water to form a solution of hypochlorite. The compound is derived from cyanuric acid, which consists of a ring of alternate carbon and nitrogen atoms with the carbons bonded externally to oxygen and the nitrogens bonded externally to hydrogen. These three hydrogens can all be replaced by chlorine to give trichloroisocyanuric acid, a white crystalline powder that is used commercially as a disinfecting agent and bleaching agent.

Alternatively, if only two of the hydrogens are replaced with chlorines, and the third hydrogen neutralized with alkali, we have NaDCC. This product is much more soluble and, in this form, it is sold as tablets for sterilizing drinking water. (They are sometimes referred to as Halazone tablets.) One of their main attractions is that they can be stored for up to three years without losing their potency, and they are more resistant to moisture than bleaching powder. The other advantage of tablets is that, unlike neat bleach, they do not damage clothing when used carelessly.

NaDCC is used in swimming pools, hospitals (to absorb and disinfect spills of blood and body fluids), animal husbandry, dishwasher tablets, lavatory-rim blocks, and floor cleaners. It is also sold in packs for travellers to take to regions where the quality of drinking water is suspect, and, while its taste may be somewhat unpleasant, at least it makes the water safe to drink. A typical tablet contains 8.5 mg of NaDCC and this will sterilize a litre of water. It forms a 2 p.p.m. hypochlorite solution, which is strong enough to kill most pathogens[29] but not so strong as to make the water undrinkable.

Hypochlorite keeps swimming pools and drinking water free from water-borne disease pathogens, not only the deadly ones, such as cholera, typhoid, and meningitis, but also the more common ones, such as *E. coli*, that cause vomiting and diarrhoea. Although the chlorination of water protects us against the ravages of microbial attack, it is not without its drawbacks, which come in the form of unwanted by-products such as chloramine (NH_2Cl) and organochlorines (for example, chloromethane CH_3Cl). These chemicals can be detected in the water of pools, and also in the air around indoor swimming pools. The chloramine arises from the reaction of hypochlorite with ammonia (from urine) and escapes into

---

[29] Such a solution is too weak to kill *Cryptosporidium*.

the air. Breathing chloramine over long periods is said to cause asthma-type symptoms in those who work in such places.

The organochlorines are formed in the water from the traces of dissolved organic matter that are present, and for this reason all water chlorination produces them, and they are even present in domestic water supplies. In swimming pools there can be twenty times as much organochlorine as in ordinary water, and much comes from hypochlorite reaction with human sweat. While this may sound alarming, it should be remembered that we are still talking about incredibly small amounts which are unlikely to affect even the very young. The chlorination of organic material uses up less than 1% of the hypochlorite, and the organochlorine compounds so formed generally have only one or two chlorine atoms added and are essentially harmless.

A level of 1 p.p.m. (0.0001%) of sodium hypochlorite will register a $\log_1$ kill (90%), and 5 p.p.m. a $\log_2$ kill (99%), of most microbes within a few minutes. It will even kill spores at a concentration of only 25 p.p.m. (0.0025%) if they are in contact with it for long enough, although a concentration of 500 p.p.m. (0.05%) will do the job much more quickly. Bleach is sold in most countries as a 5% solution, al-though, in some countries, it can be double this strength. A 5% solution has a concentration of 50,000 p.p.m., so that diluting it to the extent of using only a spoonful (10 mls) in ten litres will still produce a solution of strength 50 p.p.m., which will have the power to see off all microbes. The World Health Organization lists hypochlorite as one of the most efficient ways of dealing with HIV and hepatitis B virus.

Many modern bleaches have been thickened by the addition of polyacrylates so that they will cling more effectively to surfaces, such as that of a toilet bowl, and fragrances are now added to them to mask the hypochlorite smell. The remarkable thing about these fragrances is that they resist the strong oxidizing power of the bleach, something that would once have been thought impossible when it seemed that chlorine bleaches would remove any odour, however strong and foul, let alone the delicate fragrances of perfumes.

So what should bleach be used for? Industrially, it should be used to treat all water that is used in factories processing food; in farming, to sterilize dairy equipment; in the kitchens of restaurants and public houses; in hospitals and clinics; for the water of cooling towers; and in our homes whenever we are aware that we are at risk. Yet there are those who question the wisdom of manufacturing and using hypochlorite. Their fears are threefold: they question whether it is safe; they say it is an unnatural chemical; and they wonder whether its use might not affect sewage-treatment plants, killing the beneficial bacteria that digest human waste.

Is it safe? The answer appears to be yes, even to the extent that a child who drinks from a bottle of household bleach is hardly likely to suffer serious injury if treated promptly. Nevertheless, there are rare cases of people who have committed suicide by drinking bleach, although calculations based on experiments with rats suggest that it would require an adult to drink a litre to achieve this. (There is not enough acid in the stomach to neutralize the alkali of the bleach and thereby cause chlorine gas to form, which would be truly deadly.) Hypochlorite is not safe if it is mixed with a strong acid, such as those used to remove lime scale from lavatories. Putting both bleach and acid together will generate chlorine gas, which is to be avoided, and this is the reason why bleach is not permitted in sensitive areas such as schools.

Is it an unnatural chemical? The answer is no. In 1996, it was reported in the *Journal of Clinical Investigation* that white blood cells produce hypochlorite to fight off invading microbes. *Haloperoxidase* enzymes convert the chloride ion into hypochlorite by reacting it with peroxide, a chemical that is present in every living cell, and special white blood cells generate hypochlorite using the enzyme *myeloperoxidase* when activated by an infection. As with so many chemical discoveries that benefit us, Nature has gone before.

Does hypochlorite kill those microbes that are essential to sewage treatment? The answer is again no.[30] Once hypochlorite hits the sewers it disappears within minutes, because it has entered a strongly reducing environment and so its oxidizing power is quickly dissipated. In any case, more than 95% of the hypochlorite in bleach is transformed to chloride ions during its use. Even in countries where bleach is used daily in almost every household, with large quantities being poured into lavatory bowls, and thence directly into the sewers, there is no report of its affecting sewage plants.

[30] Where the waste from a house goes immediately to a private cesspit, then hypochlorite bleach should not be used.

Chlorinated clothing

There are cotton textiles which have an antibacterial coating that can be regenerated by rinsing with a strong oxidizing agent such as household bleach. The coating is a polymer which is bonded to cotton fabric and has chlorine atoms attached to nitrogen atoms, and it is these chlorines that act as oxidizing agents. Garments made from such fabrics work well against all kinds of microbes, and they remain effective until all the chlorine atoms have reacted. However, it is a relatively simple job to replace them by soaking the fabric in household bleach. Material of this kind has proved effective against bacteria, including the notorious antibiotic-resistant *Staphylococcus aureus* (known as MRSA, short for methicillin-resistant *Staphylococcus aureus*), as well as against fungi, yeasts, and viruses. The process for making such textiles was developed by Gang Sun of the University of California at Davis, and he claims that the treatment will survive as many as fifty washes.

(Another way of making germ-resistant clothing is discussed on page 134.)

Phenols and chloro-phenols

Antiseptic surgery introduced the benefit of phenol as a germ-killer, but it could be made even deadlier (and safer for humans to use) if the odd chlorine atom was attached.

Another chemical with a strong smell is phenol and this, too, was to play its part in preventing disease. On 12 August 1865, the Professor of Surgery at Glasgow University, Joseph Lister (1827–1912), operated on a boy whose leg had been run over by a cart, resulting in a compound fracture; this is the type of break in which the broken bone protrudes through the skin. Lister was well aware that there was a high risk of gangrene setting in; then the flesh of the leg would rot away, producing a truly obnoxious smell, and it might eventually result in the death of the patient.

Like other doctors at the time, Lister was unaware that the cause was bacterial infection—in this case by *Clostridium*, which secretes a toxin that digests the tissues—but he decided to try a new preventive treatment. He reset the broken bone and dressed the boy's open wound with lint soaked in phenol solution, a chemical that was then known as carbolic acid, and which was a by-product of the widespread production of gas from coal. No gangrene developed, the wound healed cleanly, and six weeks later the lad walked home, cured. Antiseptic surgery had begun, and henceforth was to save millions of lives. ('Antiseptic' is the term used to describe a chemical that can destroy pathogens on living tissue without damaging the tissue itself.)

It soon became apparent that, although phenol was a powerful antiseptic and disinfectant, it was sometimes used as too strong a solution, when it could damage the tissue it was meant to be protecting. By 1880, its use was being discouraged in favour of another by-product of the coal–gas industry: creosol. That industry produced lots of a brown oil known as creosote which, to begin with, was simply burnt as a liquid fuel. Later, however, it began to be used as a wood preservative, especially for railway sleepers; it was from this oil that creosol could be extracted. (*Kreosot* was the name given to a distillate from beechwood that was first produced by Carl Reichenback in Germany in 1832 and used to preserve meat, to which it gave a smoky flavour.) By 1865, other German chemists had investigated the *kreosot* from coal, and from it they obtained creosol, which turned out to be an even better antiseptic than phenol.

The phenol molecule consists of a benzene ring with a hydroxy (OH) group attached. Creosol is phenol with a methyl group also attached to the benzene ring, and this group can be next to the OH, or one removed from it, or even on the opposite side of the ring. When, in 1869, it was realized that there were these three forms of creosol, the term tricresol became an alternative name for the commercial preparations. All were equally effective at killing germs, and while creosols are also highly toxic (a tablespoonful can kill) they are not as dangerous as phenol, one gram of which has been known to be fatal. Both phenol and creosol are effective biocidal agents, and they were added to soaps to produce 'carbolic', 'coal tar', and 'medicated' soaps. They also created a demand for a new type of household product, a bottle of disinfectant.

Phenol and creosol are not particularly soluble in water, but a

PHENOLS AND CHLORO-PHENOLS

Northampton inventor, John Jeyes (1817–92), found that creosote could be made water soluble by heating it with sodium hydroxide (caustic soda) and adding the rosin of tree oils. His Jeyes Fluid was patented in 1877, and was to become a household product all over the British Empire and beyond. Jeyes did not become rich from his invention, however, because he was not a particularly good businessman, and when he died left only a modest £585.

Jeyes Fluid was ideal for use in hospitals, farms, abattoirs, food-processing plants, public lavatories, and camp-sites. It not only cleaned and disinfected but also left behind a lingering and agreeable odour. It is still made at Thetford in Norfolk, and was widely used by farmers in Britain during the 2001 foot-and-mouth outbreak. It is still popular with gardeners.

Today, these types of disinfectants come in two forms: so-called black fluid has 8% of these phenolics, while white fluid has 35%. The former is suitable for domestic use, and the latter for commercial use. When added to water, a typical black fluid would produce a white, cloudy emulsion. An alternative preparation was made from a combination of phenols and soap: it remains clear when added to water, and was marketed under the name of Lysol. This disinfectant was popular until the 1930s, when it gave way to hypochlorite bleach.[31]

Phenol may have lost its place as a common disinfectant, but worldwide more than 5 million tonnes of this chemical are still manufactured each year, and most of it goes into making polymers such as nylon and polycarbonate (see page 208). A small amount still ends up in antimicrobial products, but only after it has undergone chemical conversion to much more selective and less hazardous molecules. The way to do this is to add chlorine atoms to the phenol ring, which greatly improves its potency.

Chemically phenol is a very reactive molecule, and with chlorine water it reacts quickly to add three chlorine atoms to the benzene ring, one at either side of the phenol OH group and the other at the opposite end of the ring. The product is trichlorophenol, which is a powerful antibacterial agent and fungicide. Even one chlorine atom will suffice to boost some phenol derivatives, and benzyl-chlorophenol is one such product and this, too, is a strong germicide. Two chlorines are present in dichloro-

---

[31] Lysol is still used as a brand name for disinfectant in the United States.

phenol, and this has been used as an antiseptic but has more commonly been used to disinfect seeds. Adding four chlorines gives tetrachlorophenol, which is a wood and leather preservative, while five chlorines gives pentachlorophenol ('penta'), a strong disinfectant and fungicide, and this has also been widely used to preserve wood. Such has been the environmentalist opposition to chlorinated products, however, that many of these compounds are no longer permitted or are restricted in their use.

Hexachlorophene is still used as a germicide in soaps, handwash liquids, and lotions. It consists of two interconnected phenol rings, with each ring also having three chlorines attached. It has been particularly popular in over-the-counter treatments for acne, spots, and pimples, and is sold under names such as pHiso-Hex. In the past it was more widely used, and was added to soaps, shampoos, deodorants, and mouthwashes until the early 1970s, when tests on laboratory animals showed that it could affect the nervous system.

In theory, it is possible to make thousands of chlorinated phenol molecules by ringing the changes of chlorine atoms and other groups attached to the molecule. No doubt many were made in the last century and tested for their antimicrobial properties. Few, however, were found to combine this desirable trait with the other necessary attributes for making it a commercial success, namely complete safety for human use, ease of manufacture, miscibility with water, and compatibility with other ingredients. Nevertheless, two such products were to emerge that found wide application, and are still around today. These are *para*-chloro-*meta*-xylenol and triclosan.

Para-chloro-*meta*-xylenol is perhaps best known by its commercial name of Dettol, and it is manufactured by Reckitt Benckiser. The active ingredient is a chloro-phenol which was originally called chloroxylenol when it was first produced in Germany in 1923. Today it is better known as *para*-chloro-*meta*-xylenol (PCMX). As an antiseptic, PCMX is sixty times more potent than phenol, and it is sold as a 5% solution which includes ingredients such as a pine oil (5%), castor oil soap (14%), and isopropanol (12%), all of which help to keep it in solution.

PCMX, as Dettol, underwent clinical trials at Queen Charlotte's Maternity Hospital in London in the late 1920s, when its use as a general antiseptic cut the incidence of puerperal fever by half. Dettol is not applied neat but is diluted, and can then be safely used as an antiseptic for cleansing wounds and abrasions—it is highly effective against bacteria and fungi. Dettol was to be part of most first-aid treatments for half a century, and was also used in many other ways: added to bath water, or used as a laundry aid when washing infected fabrics, and even for the rapid disinfection of medical equipment when used in conjunction with ethanol (alcohol).

Even diluted to 1 part in 400, Dettol will kill most bacteria within five minutes and, at the recommended dilution of 1 part in 40, will deliver a log5 kill (99.999%) within one minute, even when the pathogen is present in conditions such as dirt and blood. Fungi take a little longer to kill, but again, this solution will wipe them out within two minutes. Even viruses are not able to resist its action, provided it is used at a higher concentration, such as 1 part in 10. With such a solution, herpes simplex is effectively eliminated after one minute although HIV is less affected, suffering only a log3 kill, (99.9%) in this time.

Triclosan is commonly used as a safe antibacterial agent in lots of household products, such as soaps, ointments, and toothpastes. It has also been used to impregnate plastics, especially those used as food-preparation surfaces, such as chopping boards. The Swiss-based company Ciba Speciality Chemicals (now known as Novartis) is the main manufacturer. The molecule consists of phenol linked to a benzene ring, and there are two chlorines on the benzene ring and one on the phenol.

Triclosan probably attacks bacteria at many points but, in 1998, a group at the Tufts University School of Medicine in Boston in the United States discovered that it works mainly by blocking an enzyme that bacteria need to make fatty acids; the enzyme is *enoyl-acyl-protein reductase* (also known as FabI). This means that, in theory, bacteria could evolve a triclosan-resistant strain by undergoing mutation to select an alternative enzyme to do the job, an enzyme that triclosan could not block. Indeed, one microbe, *Streptococcus pneumoniae*, has such an enzyme (FabK) that is somewhat more resistant to triclosan than FabI, and it has been suggested that this could be a route to triclosan-resistant bacteria.

What is reassuring is that, so far, bacteria show no sign of becoming triclosan-resistant and, as triclosan has been commonly used for fifty years or so, it would appear that there is really no alternative enzyme that bacteria can select to do the same job.

Quaternary ammonium salts, 'quats'

A molecule that contains a positively charged nitrogen atom to
which is attached a long hydrocarbon chain can lance through a
bacterium's cell wall and deliver it a deadly blow.

To design a chemical that will attack bacteria, three things must be incorporated into its structure: first, it should include a component that is
attracted to the bacterial cell wall and adheres to it; second, it should be
able to penetrate the wall; and third, it should be able to destroy an essential cellular component so that the bacterium will die or at least be prevented from reproducing. The chemical attack might even be on the cell
membrane itself, causing this to rupture and allowing the cell's contents
to spill out. The perfect disinfectant has yet to be designed but **quaternary ammonium salts**, the so-called quats, come close—see Glossary.

Quats meet the first of the above requirements for being an effective
bactericide, because the net positive charge they possess means they will
be attracted to the outside of a bacterial membrane which is negatively
charged. That would have little effect, however, if all the quat did was displace other positive ions that are there, such as sodium and calcium,
although this may destabilize the microbe to some extent. Once in position, the quats attack bacteria by disrupting their membranes, thereby
causing leakage which proves fatal. To be effective, the hydrocarbon
chain has to be at least 30 nanometres long. This acts rather like a sword
which stabs into the membrane and simply ruptures it in a way that cannot be repaired.

The German biochemist Gerhard Domagk (1895–1964) set the stage
for the discovery of antibacterial quats while he was working on the
sulfonamides; these were among the first successful antibiotics. They
are amine-type compounds and their nitrogen atoms can easily become
positively charged. Other compounds also interested Domagk, however,
and he discovered that the particular feature that guaranteed antibacterial activity was having a hydrocarbon chain with between 8 and 18 carbon
atoms attached to the nitrogen. While many such compounds have since
been tested, only a few have been commercially successful.

The best of the quats, and the ones that have withstood the test of time, are benzalkonium chloride and cetrimide, both introduced in the 1960s. Benzalkonium chloride has a multitude of trade names, such as Sephiran and Germinol, and it consists of a nitrogen atom to which are attached two methyl groups (CH_3) and a benzyl group (that's a methyl group with a benzene ring attached). The fourth bond to nitrogen is to the hydrocarbon chain of between 8 and 18 carbon atoms, and, as its name implies, the balancing negative charge is a chloride ion, Cl^-. Benzalkonium chloride is very soluble in water, which is what is required if such a material is to be of any use.

Cetrimide is a mixture of quaternary ammonium salts in which the central nitrogen has three methyl groups and one long-chain hydrocarbon group, which can be 12, 14, or 16 carbon atoms long. It was first made in 1946, is still used in antiseptic skin ointments (of which Savlon is the best known), and has remarkable skin-healing properties.

There are times during an epidemic when it would be reassuring if all public surfaces that people handle, such as doorknobs, switches, counters, chairs, menus, and so on, were germ-free. Perhaps one day they will be, and that might come about through the use of quats that are chemically bonded to such surfaces, thereby making them permanently deadly to bacteria. Such products are now being manufactured.

Bacteria with which a normal healthy person can live can be a threat to those whose immune system is impaired, or who are badly burned. Such individuals tend to be isolated in intensive-care wards where every effort is made to prevent secondary infections reaching them. And yet, despite the best efforts of doctors and nurses, some do succumb to antibiotic-resistant bacteria, the so-called superbugs. In 1996, it was reported in the journal *Annals of Internal Medicine* that such bacteria can survive for long periods on the uniforms of healthcare workers, suggesting that clothing might be a vector for spreading disease. Just as textiles can be modified with various chemicals directly bonded to their fibres, such as dyes, UV filters (see page 25), and flame retardants, so there is no reason why biocidal molecules could not be so attached. In the past, fabrics have been produced that have been treated with anti-microbial agents, although such treatments were not permanent and the agent would eventually be laundered out. Now, however, there is the possibility of making the treatments an integral and permanent feature that is capable of with-standing repeated washing.

A fibre containing bound quat molecules was first introduced in the early 1970s by the US chemical company Dow; it has been widely used in Japan, China, and other Asian countries where the aim was to counter bacteria that produce unpleasant odours. More recently, a group of chemists based at the Massachusetts Institute of Technology in the United States has found a way of bonding quats to all kinds of material, such as glass, polyethylene, polypropylene, nylon, and polyester, and these have then been proved to be effective against bacteria.

Meanwhile, other American chemists have found it possible to bond quats to carbohydrate-based materials such as cotton cloth, wood, and paper, and these quats carry a 16-carbon hydrocarbon sword. Such car-bohydrate-treated materials are thought to be potentially useful for sur-gical dressings and medical textiles, and these could be washed without washing out the quats. Tests have shown that such materials prevent the growth of many bacterial strains.

The German company Degussa has developed a polymer based on polyamines that is active against a range of bacteria including *E. coli* and *Staphylococcus aureus*, as well as against yeasts, fungi, and algae. What is unusual about the new polymer is that it is active, whereas the monomer from which it is made has no anti-microbial activity. In the polymer, the amino groups which carry a partial negative charge are concentrated along the polymer chain and they protrude from its helical structure. These charged polymer chains are able to interfere with the micro-organ-isms they come into contact with, and have been proved to be effective at killing Gram-positive and Gram-negative bacteria, as well as moulds, yeasts, and algae. Possible uses for the new material are in wood pre-servation, in anti-fouling paints for boats, in food-processing plants, in water-treatment plants, and to protect historical artefacts.

Hydrogen peroxide

Regarded as one of the most environmentally friendly of chemicals, hydrogen peroxide is nevertheless deadly to bacteria as well as being a gentle bleaching agent. We benefit from it today because of the Nazis' obsession with destroying London using long-range missiles.

Hydrogen peroxide (H_2O_2) is an excellent oxidizing agent, and it will destroy most microbes. In so doing, it produces as a by-product only the environmentally friendly hydrogen monoxide (otherwise known as water, or H_2O). The trouble with hydrogen peroxide is that it is intrinsically unstable, with a tendency to decompose to water and oxygen unless precautions are taken. This decomposition is catalysed by traces of metals, dust, and enzymes. One particular enzyme, *catalase,* specifically targets hydrogen peroxide, and a single enzyme can consume 50,000 hydrogen peroxide molecules *per second.* You can see this rapid decomposition happening before your very eyes if you add a piece of uncooked mince to hydrogen peroxide. (You could use grated raw potato, but the process is slower.)

Plants and animals need *catalase,* because hydrogen peroxide is part of their metabolism—our own body generates around 30 grams of hydrogen peroxide per day—but it can be a threat because it is a potential source of dangerous free radicals. Other species can produce H_2O_2 at a relatively much higher rate, because they use it to generate light by a process known as chemiluminescence, and, in the inky black depths of the oceans, some fishes attract their prey this way.

In the home we can use hydrogen peroxide for disinfecting and for bleaching. For almost a century it was used as an antiseptic and a hair decolourant, and it still is. It can also be used to remove oxidizable stains on garments, table linen, sofas, and carpets; it works moderately well, and is safer than hypochlorite bleach in that it will not decolourize the fabric that is stained. Such products contain about 8% H_2O_2.

Hydrogen peroxide can also be purchased as a dry powder in the form of sodium percarbonate, which can be added to the laundry to boost cleaning, or mixed with water and applied as a slurry to clean stained

> **Dazzle them with a peroxide smile**
>
> Hydrogen peroxide gels are good at whitening teeth, and they can be applied either as a paste or as an adhesive strip. Within 15 minutes of being applied, the H_2O_2 diffuses through the tooth enamel and into the dentine, which is where the molecules that cause the discolouration of teeth are to be found. There the peroxide bleaches them, and the natural whiteness of the teeth re-establishes itself, although it will take several applications to achieved the desired Julia Roberts smile.
>
> The H_2O_2 comes in the form of urea–peroxide, a combination of urea and H_2O_2 held together by **hydrogen bonds** (see Glossary). Also present in the gels are other chemicals, some that form the gel, some that give it a pleasant flavour, and some that function as preservatives to prevent decomposition of the H_2O_2.

surfaces. The kinds of stains that peroxide is best at removing are those of coffee, tea, red wine, and fruit juices. Hydrogen peroxide is also good at disinfecting things, and the slurry is particularly good at removing microbial growths from patios and outdoor wooden furniture. H_2O_2 will keep swimming pools and hot tubs free from bacteria, provided it is maintained at a concentration of about 100 p.p.m.—which is a much higher level than that of hypochlorite needed to keep such pools and tubs free from pathogens. Hydrogen peroxide can be used as an antiseptic wash for cuts and abrasions, as a mouthwash, in toothpaste, to get rid of athlete's foot, and to treat ear infections.

A new method of sterilizing rooms employs hydrogen peroxide as a vapour, and is known as VPHP (vapour phase hydrogen peroxide). It is preferable to the more usual gaseous sterilants, formaldehyde and ethylene oxide, which are toxic and dangerous to handle, although even VPHP-sterilized areas can be dangerous to enter if the level of residual H_2O_2 is too high. Special equipment can turn reagent-grade H_2O_2 (30% strength) into vapour with a concentration of around 1–2 mg per litre of air at 25 °C, and this is blown into chambers and machinery where conditions must remain sterile, as in the packing of pharmaceutical products. VPHP was introduced in 1991 and has now been installed in more than 500 locations around the world.

Hydrogen peroxide has a long history; it was discovered in 1818 by the French chemist Louis Thénard,[32] who reacted barium peroxide (BaO_2) with sulfuric acid. (He made his barium peroxide from barium oxide (BaO) simply by heating it in air.) He studied the new material for many years, even obtaining an almost pure sample. That was quite an achievement considering its instability, which, unknown at the time, was the result of tiny amounts of dissolved ion impurities of metals such as iron and manganese that catalyse its decomposition. Indeed, commercial production of hydrogen peroxide by Thénard's method began in only 1873, in Berlin, and the product had to be used almost as soon as it had been made.

Even when an electrochemical method of making H_2O_2 was introduced in the early twentieth century, based on the electrolysis of pure sulfuric acid or potassium sulfate, the product still had a relatively short shelf-life of only a few weeks. Nevertheless, industrial-strength hydrogen peroxide, of 30% concentration, was manufactured and used for bleaching, and domestic grades of 3% and 6% were sold for use in the home. The biggest boost to H_2O_2 production came from the Third Reich's military programme of the 1930s and '40s.

Nazi scientists and engineers used H_2O_2 for the world's first liquid-fuelled rocket engine for aircraft, first tested in 1936. A rocket-powered fighter, the Komet, even entered service towards the end of World War II. It was powered by H_2O_2 reacting with a mixture of hydrazine and methanol, and it could reach speeds of 965 km/h. Hydrogen peroxide was also a necessary part of the world's first ballistic missile, the V2, which became operational in September 1944; it was directed mainly at London, and then at Antwerp after that port had been recaptured by the Allied forces.

Hydrogen peroxide was also needed to power the pistons of the catapults that launched the most destructive of the German 'wonder weapons', the V1, also known as the flying bomb or 'doodle-bug'. These were

[32] Thénard (1777–1857) came from poor peasant stock, and he washed bottles to earn his keep while studying chemistry at the École Polytechnique in Paris. He soon showed himself to be a brilliant chemist and was made a professor when only twenty-seven. He discovered a bright-blue pigment which became known as Thénard's blue, which was much used in porcelain in the nineteenth century. Later in life he went into politics and served in France's Chamber of Deputies.

the prototypes for today's cruise missile, and were designed to devastate London, where they did, indeed, inflict enormous material damage. Although the V1's guidance system was primitive, and its time-of-flight determined by a set number of revolutions of a small counter, it was a terrible weapon. Of the 5000 or so aimed at London, 2419 reached that hapless city between June 1944 and March 1945, some causing dreadful damage.[33] The V1 was launched up a ramp by means of a piston catapult that propelled it into the air on a blast of oxygen and steam, generated from the reaction of 100 kilograms of hydrogen peroxide and potassium permanganate.

The hydrogen peroxide had to be highly concentrated (80% H_2O_2) and, to produce this, German chemists devised a method based on the reaction of an anthraquinone, first with hydrogen gas, H_2, and then with oxygen gas, O_2, the overall reaction being $H_2 + O_2 = H_2O_2$.[34] The anthraquinone was then regenerated and the process repeated again and again. The method was so efficient that it was to become the modern method that is used today.

The use of hydrogen peroxide in warfare is not a human invention; the bombardier beetle perfected a similar launching method several million years ago. This insect, *Stenaptinus insignis*, is to be found in tropical countries, such as Kenya and Malaysia, and is about 2 cm long. When threatened, it shoots rapid bursts of a boiling hot spray of an irritant, quinol, from a nozzle situated at its rear end, which it can direct at its enemy. The pressure to emit the spray of liquid, and its high temperature, is generated from the catalytic decomposition of hydrogen peroxide by *catalase*. The quinol and hydrogen peroxide are kept in one sac inside the beetle and the enzyme fluid in another, separated by a valve that the beetle opens when it feels threatened. The insect can discharge up to thirty shots before it runs out of ammunition, and, provided it escapes its predator, can replenish its stock of hydrogen peroxide within a day.

Hydrogen peroxide production soared after World War II, albeit for

[33] The worst incident was the V1 that struck the Guards Chapel near Buckingham Palace during a packed service on Sunday, 18 June, killing 120 worshippers, mainly servicemen, and badly injuring about 150 others.

[34] In the 1990s, this reaction was performed with ozone (O_3) instead of oxygen, and the product was the unexpected molecule dihydrogen trioxide, H_2O_3, in which there are three linked oxygen atoms with a hydrogen at each end, i.e. H–O–O–O–H. It is much less stable than hydrogen peroxide, H–O–O–H.

less dramatic roles than powering aircraft and missiles, and launching weapons. The older uses of bleaching linen, cotton, and wood pulp were to be joined by other functions, such as the manufacture of sodium perborate, the bleaching agent once widely used in laundry detergents. To a large extent, this compound has now been replaced by sodium percarbonate, which is simply a combination of sodium carbonate and hydrogen peroxide in the ratio of 1 to 1.5.

The hydrogen peroxide made today comes from the anthraquinone process, which produces an aqueous solution of between 20 and 40%. This is then concentrated by vacuum distillation to commercial grades of 50 to 70%. Vacuum distillation is carried out under much-reduced pressure, which means that the water can be removed at temperatures lower than its normal boiling point of 100 °C, while the hydrogen peroxide, which normally boils at 155 °C, remains behind. The final product is stored and transported in tanks made from stainless steel or aluminium. By adding stabilizers to hydrogen peroxide, it is possible to extend its shelf-life to such an extent that less than 0.1% per month is lost to decomposition. This protection is afforded by complexing agents, such as sodium stannate and various phosphates, which bind themselves to any metals that are present, thereby making these incapable of catalysing the decomposition.

More than a million tonnes a year of hydrogen peroxide are manufactured world-wide. About 30% of this goes into pulp and paper bleaching, and 20% into textile bleaching, with a similar amount converted to the more stable solid peroxides. The remaining 30% is used in many ways in industry: as a chemical resource; as a disinfectant for cleaning and deodorizing polluted water; for separating metals from their ores, for example in the extraction and purification of uranium; and for laundry and dishwasher products. (Relatively little is sold to the public in its traditional liquid form.) Hydrogen peroxide is also used to manufacture the chloro-phenol PCMX mentioned above.

Hydrogen peroxide has other valuable properties apart from its ability to bleach and to kill germs. In an emergency, it could vastly increase food supplies by making otherwise inedible wastes, such as straw and sawdust, give up their cellulose by freeing this from the indigestible lignin to which it is chemically bonded. The treated material could then be used as fodder for ruminants such as cows which are able to digest cellulose. The trick is to use an alkaline solution of 1% hydrogen peroxide at a pH of between 11 and 12. Nature, too, uses this technique, and fungi that live off wood (and even mushrooms) secrete their own alkaline hydrogen peroxide.

Although, in theory, hydrogen peroxide is a strong oxidizing agent, it needs to be activated in order to work. In industry this is done with an acid, such as sulfuric acid (H_2SO_4), which is then converted to peroxysulfuric acid (H_2SO_5), known as Caro's acid, or with acetic acid (CH_3CO_2H), which is converted to peroxyacetic acid (CH_3CO_3H), and it is this that performs the actual oxidation. Alternatively, hydrogen peroxide can be activated by an alkali, or by a catalyst, or even by UV light. We can think of these various processes as breaking H_2O_2 into more active fragments, such as HO^+ and HO_2^- ions, or $HO\cdot$ free radicals (the dot signifies the lone electron).

The chemical industry likes hydrogen peroxide because the end products of its use can only be water and oxygen. Such is the aura of environmental approval of hydrogen peroxide that it now has halo status—in other words, it is seen as a godsend by those who disapprove of 'chemicals' per se, but who still wish to enjoy the benefits of the chemical industry.[35] Indeed, its adherents have widened its use way beyond that which it can truthfully deliver, and there is no scientific evidence to support the claims they make.

Adding a cupful of hydrogen peroxide to your bath water and soaking in it for thirty minutes is said to be rejuvenating and detoxifying. Dabbing your face with a 3% solution is said to get rid of acne. Spraying yourself with such a solution after you've showered is said to tone up your skin, and colonic irrigation with it likewise tones up your intestines and bowel. Watering plants with a solution made up of 1 part of the 3% solution and 40 parts of water is said to be good for them, and seeds germinated by soaking in such a solution are believed to produce better plants. Chickens reared on water containing a few drops of hydrogen peroxide are said to lay tastier eggs, and are themselves tastier to eat when they've stopped laying; and, when cows are given it, they produce more milk. All these claims should be taken with a pinch of salt, though such uses probably do little harm.

[35] It is literally a godsend because it is present in rainwater, probably derived from ozone.

One use appears to be even more ridiculous, however. Drinking a hydrogen peroxide solution, heavily diluted with spring water (naturally), is said to prevent or treat all kinds of diseases from multiple sclerosis to AIDS, not to mention cancer, arthritis, and asthma. The Church of Zion, which has several thousand adherents in Canada and Hong Kong, used to exhort its followers to take a daily shot of the cure-all. (Readers should be warned that when hydrogen peroxide is first tasted it seems somewhat sweetish but, within a few minutes, a bitter aftertaste develops.) Those who engage in this dubious practice need to be warned of its unpleasant side effects. 'Cleansing' the body with hydrogen peroxide can lead to nausea, headaches, fatigue, boils, and diarrhoea—all of which are misleadingly said to be part of the body's toxic elimination process.

Ozone

This may be the gas that protects the Earth from the deadly rays of the Sun but, as a chemical, it too is deadly, not only to germs but also to humans as well, yet we benefit from its use in things as diverse as bottled water, swimming pools, and sewage.

In 2001 the Food and Drugs Administration in the United States approved the use of ozone as an anti-microbial agent—but only in certain situations that don't involve exposure of the public to this toxic gas. A hundred years ago, people were much more enamoured of ozone: indeed, so convinced were they that it killed germs, ozone generators were installed in churches and public halls, and travellers on the London Underground were 'protected' with it. In an age when it was believed that most germs were spread by the air we breathe, then, clearly, a powerful oxidizing gas that could kill them was surely the best method of dealing with the threat.

Medical commentators were even able to back up this theory by reference to the better health of those who lived in areas where the level of ozone was erroneously assumed to be high, such as by the sea or in mountainous regions. Of course, the theory was wrong, and we now know that breathing ozone gas can only damage the lungs. Yet today-more ozone is manufactured than ever before, although almost all of this is used to disinfect water, from the ultra-pure water used in pharmaceutical manufacturing to that discharged into rivers from sewage works.

Ozone gas in the atmosphere occurs in the upper and in the lower regions. High up, it functions as a protective layer that screens out the dangerous UV radiation from the Sun. Down here at the surface, it is a pollutant, damaging both plants and animals. It used to be produced mainly from the exhaust gases of automobiles in the days before catalytic converters. While the engines of older cars did not emit ozone as such, they did emit nitrogen oxides and especially nitrogen dioxide (NO_2). This can be decomposed by strong sunlight to nitric oxide and oxygen atoms ($NO + O$), and the latter then went off to react with ordinary oxygen (O_2) to form ozone (O_3). Thus, on hot sunny days, the level of ozone in cities and their surroundings could be as high as 0.1 p.p.m., which is the legal limit for occupational exposure, and even levels as high as 0.25 p.p.m. have been recorded. The natural background level of ozone, in the air is only 0.02 p.p.m., which is safe. Components of the atmosphere, such as dust, catalyse its decomposition back to ordinary oxygen.

Ozone is soluble in water to the extent of 570 p.p.m. and, while such a solution is not very stable, it is nevertheless powerfully germicidal. The water of public swimming pools is often sterilized with ozone, although a little hypochlorite still has to be added to the water that has been sanitized because surplus ozone is removed before the water is returned to the pool.[36] In any case, it would not persist long enough to offer the best protection to bathers. This is the reason why the ozonization of public water supplies would not be practicable in many areas: by the time the water had travelled to the consumer, all the ozone would have disappeared and the protection it offers would have been lost. Nevertheless, in some countries, such as France, ozonization has been used for more than a century (the first such plant was installed in the Netherlands in 1893), and is the preferred way of ensuring the safety of public water supplies. Ozone is an even stronger oxidizing agent than hypo-chlorite, to the extent that it can wipe out *Cryptosporidium*, which is somewhat resistant to very dilute hypochlorite. Ozonization is becoming more popular, especially where people worry about chlorination. An ozonization plant

[36] This is done by passing the water through a charcoal filter, which catalyses the decomposition of ozone to oxygen.

to treat drinking water in California was installed in 1987 but, within ten years, more than 250 such plants were in operation in that state.

Those who drink bottled water can also thank ozone for ensuring its 'purity'. While carbonated water is protected by its acidity, non-carbonated water is susceptible to microbial growth. The best way of dealing with this is to ozonize the water, which will then remain germ-free once the bottle has been sealed. Ozone also protects us by disinfecting the water of cooling towers and by treating that which is discharged from industry. Industrial effluent may be contaminated in many ways, and oxidizing it with ozone not only removes microbial contamination but also other types of pollution as well. Ozone for such processes is made on site and used immediately it is formed.

The water for aquariums, fish farms, fish hatcheries, and shrimp farms is better protected by ozonization than by chlorination. Ozone generators are easily installed, and the treated water can be de-ozonized by passing it through a charcoal filter before it is returned to the fish and shrimp tanks. The gas is also being used to clean up waste water from sewage, especially if it is to be discharged near holiday beaches. It not only kills bacteria but also kills the obnoxious smell of such water by oxidizing odorous sulfides, such as hydrogen sulfide, to odourless sulfate.

Ozone was first identified by the German chemist Christian Friedrich Schönbein in 1840, at the University of Basel in Switzerland, and he named it for the Greek word, *ozon*, meaning 'to smell'. In a way, its history goes right back to the ancient Greeks, who noted the strange smell that sometimes accompanies a thunderstorm, and, indeed, raised levels of ozone are produced by lightning. At first, Schönbein thought that he had discovered a new element similar to chlorine, but eventually he proved that it was simply another form of oxygen with three oxygen atoms joined together instead of the usual two.

Ozone can be produced in two ways: the best way is to pass air through concentric glass tubes with metallized surfaces and apply a corona discharge voltage of 15 kilovolts between them. The air emerging from such a tube will have 2% of ozone, and there are generators that can produce up to 500 grams of ozone per hour, although many small-scale applications will require only 1 to 2 grams an hour. The other method of ozone generation is to subject air to ultraviolet light, and this will be sufficient if only low concentrations of the gas are needed. It is possible to condense ozone out of a stream of air if it is cooled to below its boiling point of $-112\,°C$, and this yields a blue liquid which, on cooling to $-193\,°C$, will form a violet-black solid. Both liquid and solid are dangerously explosive.

Super-germs?

Some people worry that overuse of anti-microbial chemicals will eventually lead to resistant strains that may be even deadlier than those we live with at present. Whether these fears are well founded is debatable. Chemical reagents that are powerful germicides don't just attack a microbe at one vulnerable point, say a key enzyme, a situation which it might be able to resolve by developing a modified enzyme that is not vulnerable to the attacking agent. That certainly has happened with antibiotics, but by their very nature these chemicals have to be highly selective in how they attack, otherwise they might inflict similar damage on living cells of the human body. On the other hand, disinfectants are more like fragmentation bombs, scattering their chemical missiles at all kinds of vulnerable points in the microbe, be they enzymes, RNA, membrane molecules, or messenger molecules. Furthermore they act so rapidly and are so deadly that microbes stand little chance of developing the multiple strategies needed to defeat them.

Shower cleaners

The secret of shower cleaners is to make water wetter, and better, at dissolving things.

I began this chapter with an ironic comment about modern kitchens, how clean they were and how little they are used for their traditional purpose of preparing foods from raw vegetables and basic ingredients. In fact, many people these days probably spend more time in their bathrooms than in their kitchens, which is no doubt why the next product was such an instant hit. The curious thing is that it wasn't designed to make showers germ-free zones although undoubtedly it helps, by making them so much easier to clean.

Provided a shower is clean to begin with, it is possible to keep it forever clean. All you need to do is spray the walls with one of the new shower cleaners after you've had your shower. No rubbing, no rinsing, no wiping, no worry. The first shower cleaner, called Clean Shower, was formulated by Robert Black in Jacksonville, Florida, in 1997, and it came about when Robert's wife asked him to clean the shower one day. He found it such a chore that he decided there must be a better way than using conventional cleaners. The upshot was US patent 5,910,474. Its principal ingredients were a surfactant, a chelating agent, and a solvent. A typical recipe, as given in the patent, consisted of an ethylene glycol non-ionic surfactant (1.5%), diammonium ethylenediamine tetracetate (1.5%), and isopropyl alcohol (4%), plus a few drops of fragrance. Clean Shower contans a non-ionic surfactant called Antarox BL-225, which is manufactured by the chemical company Rhodia, and it was chosen because of its excellent cleaning and solubilizing properties.

Other companies were quick to bring out 'me-too' versions of Clean Shower, some of which were not particularly good, but all of which relied on a similar combination of active ingredients, each of which has a role to play. A good shower cleaner needs an antibacterial agent added to it, such as a quat, otherwise it may become contaminated so that you end up spraying the shower with germs as well as cleaning agents. In 1999, a quarter-of-a-million bottles of one brand of shower cleaner sold in the United States had to be recalled by the manufacturer because they were so contaminated.

Surfactants come in three main types: cationic, anionic, and non-ionic.[37] The first of these has a positively charged group at its head (cations are positively charged ions); the second has a negative group (anions are negatively charged); and the third category has neither. The surfactant has the job of making the water more wetting. In scientific terms, it reduces the surface tension of water droplets to the extent that they collapse to form a film of water, which simply runs off the various surfaces, an effect known as sheeting. Normally, water will remain as drops on the surfaces of the shower; these evaporate and leave a mark caused by the things they contain, such as dissolved salts, body oils, and the re-mains of skin lotions and shower gels.

The chelating agent in the shower cleaner is there to dissolve metal

---

[37] Cation is pronounced 'cat-ion', and anion is pronounced 'an-ion'.

ions such as calcium and magnesium which may be present in the water supply, these being the metals that account for the hardness of water and the formation of soap scum. The best way of keeping these metals soluble is to add **EDTA** (see Glossary). This molecule is ideally structured to wrap itself round calcium and magnesium ions, making it impossible for them to precipitate out of solution as they are inclined to do.

The solvent in a shower cleaner is there to solubilize traces of the various oils that come from the human body, or from the skin lotions that we have put on it. Robert Black chose isopropyl alcohol (more correctly called 2-propanol) as a safe solvent; it has been used in body lotions and after-shave lotions for many years, and is generally regarded as safe.

The usual way of removing oil and grease is to attack it chemically using a cleaning agent that is alkaline, with a pH of about 12 or so, because this breaks down the oil and grease into glycerol and fatty acid ions which are soluble in water and can be washed away. Obviously, in the close confinement of a shower the user will be standing naked, and a cleaning spray that is an alkaline solution has to be avoided. For this reason, shower cleaners have a pH of around 5 (which is the same as rainwater) and, to keep them at such a pH, they may be **buffered** (see Glossary) with sodium citrate and citric acid .

Once you've sprayed inside the shower with a shower cleaner, it will leave the walls apparently clean, although there is bound to be a little residue of the shower cleaner's ingredients, but these then act as a barrier for when the shower is next used. Of course, if your shower is already stained with age, and the grouting between the tiles is stained with mildew, then you will have to give it a thorough cleaning before incorporating a shower cleaner into your toilet routine.[38]

---

[38] Mildew tends to form in the grouting because grout holds water long enough for the mould to grow. The best agent for killing this is hypochlorite bleach, in particular thickened bleach that will remain in contact with the grout until the mildew has all been removed. Special grouts and sealants are now available with built-in anti-mildew agents.

CHAPTER FIVE

It's All in the Mind

I N THIS CHAPTER, we shall look at the effects of chemicals on the brain. We shall look first at the drugs that are used to treat common forms of depression, then at lithium, which is used to treat a much more serious mental condition, and finally at aluminium, which has been accused of causing the brain-destroying illness Alzheimer's disease. Before we launch into the chemistry of the molecules that can ease the troubled mind, however, we shall look first at the condition that probably affects most people at some time in their lives: depression.

The human brain

The average brain weighs 1.4 kilograms (3 pounds) and floats within the skull in cerebrospinal fluid. It consists of an outer layer, known as grey matter, which is the cerebral cortex, and an inner layer, which is white, with cavities known as ventricles. The brain is fed by large arteries that pump 1.7 litres (3 pints) of blood to it every minute, providing it with the oxygen and glucose it needs. Indeed, the brain demands 20% of the body's oxygen intake.

At birth, the brain has 100 billion nerve cells (neurons) each of which can have between 1000 and 10,000 connections to other neurons, making a total of 100 trillion links—far in excess of even the world's most advanced computer. After birth, and from then on, the brain loses about half-a-million neurons a day, which sounds rather frightening but is small in percentage terms (0.0005%), although this loss accounts for about 10% of the brain's capacity during a normal lifetime.

Depression and antidepressants

It has been estimated that 5% of the population is
suffering from depression at any one time, and many now turn to
their doctors for help. They have a range of products that can
provide various kinds of relief, depending on the type of
depression from which the person is suffering.

Anyone can be affected by depression, including the rich, the famous, and the powerful. Indeed, the depressions of such people as Winston Churchill, Charles Dickens, and Florence Nightingale are well documented. Nor is depression just a medical condition of advanced societies. We are sometimes led to believe so, and we imagine that, in earlier ages people lived simpler lives, free from the stresses of debt, job insecurity, loss of possessions, competitiveness, and relationship breakdown. We all get depressed at times when things go wrong, but that's part of life, and as circumstances change we know our depression will go away. Meanwhile, we take measures to cheer ourselves up, such as having a drink, smoking a cigarette, eating a bar of chocolate, or going out for an evening with friends.

Clinical depression, on the other hand, is very different, and traumatic events in our life can precipitate it. Mostly its cause is impossible to determine, and its effects are beyond our control because it stems from the way our brain handles the essential messenger molecules by which it functions. These chemicals, called neurotransmitters, are a key part of the brain's make-up, and there may be too few of them—or too many. One woman in five and one man in ten are likely to suffer from clinical depression at some time in their lives and, once you've had one such episode, which can last up to two years, the chances are that others will follow.

Ancient Greek and Roman physicians referred to conditions we would now recognize as clinical depression. In the fourth century BC, Hippocrates described it while, in the second century AD, the great Roman physician Galen attributed it to an excess of 'black bile', one of the four 'humours' that were supposed to govern our health. In the Middle Ages, people talked of melancholy, using this as a term for unexplained sadness, and when Shakespeare uses the word in *Twelfth Night* (Act II, Scene iv) he seems to be describing clinical depression:

> VIOLA   . . . She pined in thought,
> And with a green and yellow melancholy
> She sat like patience on a monument,
> Smiling at grief . . .

Today, we recognize several kinds of depression, of the types that cause a person to seek medical help rather than suffer in silence. There is the depression caused by a specific event or set of circumstances; this used to be known as reactive depression, and it is very different from the depression with no obvious cause, which was known as endogenous depression. The former is now called unipolar depression, meaning that the person concerned has been affected by an overwhelming life event, such as the death of a partner, the break-up of a relationship, failed examinations, or the loss of a job. (A more protracted form of depression is called dysthymia. This lasts for two years or more, and is present for some time during most days.) The disorder in which the sufferer swings between days of deep depression and those of unnatural elation, previously known as manic-depressive illness, is now called bipolar depression. There are other forms of depression, such as post-natal depression, seasonal affected disorder (SAD), which occurs in late autumn and winter, and the secondary depression that comes as part of other diseases, such as diabetes, coronary heart disease, arthritis, AIDS, cancer, and stroke.

Effective medical treatments for such mental states first became available in about 1900 when 'sedatives' first appeared. These were derivatives of barbituric acid, a simple chemical that was first made in the late nineteenth century by the German chemist Adolf von Bayer. By the year 1900, more than 2000 barbiturate variants had been produced and tested, and some of them, such as phenobarbital (also known as phenobarbitone), became the standard treatments for depression. One barbiturate, known as Seconal (secobarbital sodium), was used as a treatment for insomnia. Another, Pentothal (thiopental sodium), became known as the truth drug because it was believed to make people under interrogation give information more readily. The barbiturates are now rarely used because an overdose can kill, and such was the danger they posed that warnings were broadcast on national radio whenever phenobarbitone tablets had been lost or stolen.

The medical profession also resorted to other methods to treat severe depression, such as electroconvulsive therapy in which a massive electric shock is applied to the brain. This so affects the neurotransmitters that their effects became muted, and this offered relief.

Better antidepressants were discovered by accident. Amphetamine, for example, was originally prescribed in the 1930s as a nasal decongestant, but had the side effect of insomnia. This drug, known by the name Benzedrine, was also given to bomber crews in World War II to keep them awake on long missions. We now know that the main effect of amphetamine is to cause a sudden rush of two messenger molecules, dopamine and noradrenaline, while suppressing the effects of a third neurotransmitter, serotonin. Amphetamine's action on serotonin also suppresses the appetite, and this is why it is popular among dieters, while its action on noradrenaline makes it a favourite among students studying for exams, as well as soldiers going into battle.

By the 1950s, it became possible for doctors to prescribe less aggressive and safer treatments. The first of the modern antidepressant drugs was iproniazid, which was originally designed as a treatment for tuberculosis. Doctors who used it, however, began to report how it greatly improved the mental state of their patients. They also reported that another drug, resperine, had just the opposite effect. Designed to alleviate high blood pressure, it also made patients acutely depressed. This spurred researchers into investigating the mechanisms by which the brain operated, and, in the case of resperine, they discovered that it was depleting the brain's store of messenger molecules.

Pharmaceutical companies began to research this area and they came up with drugs that could affect brain chemistry in a positive way. One of the first to be marketed was Tofranil (generic name imipramine), which consisted of three rings of atoms. It was so effective that it was joined by similar drugs based on the same three-ring structure, and these became known as the tricyclic antidepressants. Other tranquillizers, as they were now being called, were based on a two-ring structure, and some of these, such as Valium (generic name diazepam) and Librium (generic name chlordiazepoxide), became widely prescribed. They both work by being converted in the body to another form of the drug called nordazepam. The sleeping pill Mogadon (generic name nitrazepam) is also of this type. These drugs have been largely superseded, although Valium is still used by surgeons as a muscle relaxant.

Clinical depression can be identified by a set of symptoms that must be present for at least two weeks. These are a feeling of misery and sadness, lack of interest in activities that would normally give pleasure, lethargy, sleeplessness, loss of appetite, continual anxiety, and even physical symptoms such as rapid pulse, headaches, stomach pains, and loss of weight. Together they are likely to lead to a doctor prescribing a modern antidepressant drug, and the best known of these are Prozac, Zoloft, and Paxil; of these, Prozac has become the most popular. These drugs affect the chemistry of the brain in ways that restore the natural levels of neurotransmitters.

The brain's billions of neurons have axons extending from them along which messages are passed. The cells, also put out filaments, known as dendrites, which connect cells, but it is the axons along which messages are passed. Somewhat strangely, however, axons do not link one brain cell directly to another. Along the route there is a break, called a synapse, and it is at this juncture that the messenger molecules operate. They are released from the end of the neuron, pass across the synapse gap, and then trigger a receptor which carries the message forward. There are many such messenger molecules (for example, nitric oxide, which was dealt with in Chapter 3), but three have been identified as particularly important in determining our mood; they are dopamine (also called 3-hydroxytyramine), noradrenaline (also called norepinephrine), and serotonin (also called 5-hydroxytryptamine).[39]

Dopamine is produced by our body from the amino acid tyrosine, and noradrenaline is formed from dopamine, while serotonin is made from another amino acid, tryptophan. Chemically these messenger molecules are all amines, meaning they have active nitrogen centres, and they are referred to as biogenic amines because they arise within the body.

Although we cannot directly link these brain chemicals to specific

---

[39] Other neurotransmitters are acetylcholine, glutamate, and gamma-aminobutyric acid (GABA).

mental states, there are some indications as to which emotions they are involved in.

Neurotransmitters and emotions

Neurotransmitter	Positive effects	Feelings resulting from a lack of neurotransmitter
Dopamine	General feeling of well-being	Withdrawn and inhibited
Noradrenaline	Energy, vigilance, motivation	Rejected and dejected
Serotonin	Feeling of security and happiness	Vulnerable and hungry

---

Serotonin itself has several roles in the body and is associated with the pattern of waking and sleeping, as well as controlling the appetite, and with feelings of comfort, security, and confidence. Studies have shown that prisoners generally have low serotonin levels, as do those children showing antisocial behaviour. Serotonin acts as a brake on impulsive behaviour, and individuals who have high levels of serotonin tend to think before they act, whereas those with naturally low levels tend to act first and think later, a course of action that often lands them in trouble. Monkeys with low serotonin levels generally bear the most scars on their bodies, evidence of their tendency to resort to violence.

Neurotransmitters are released from the transmitting axon into the synapse gap, and it is there that things can go wrong. After a messenger molecule has delivered its message, it can be deactivated by an enzyme known as *monoamine oxidase,* or it can be reabsorbed into the transmitting axon, ready to be used again, a process known as re-uptake. Either process will reduce the level of the neurotransmitter molecule in the synapse gap, to such an extent that it may interfere with the molecule's allotted task, resulting in mood changes and depression.

There are two ways of dealing with the lack of messenger molecules: either block the activity of the *monoamine oxidase* enzymes or obstruct the re-uptake process. The former is likely to affect all three neurotransmitters, whereas the latter approach can be more carefully targeted because each messenger molecule has a specific re-uptake mechanism. In theory, therefore, it should be possible to exert finer control over the one that is suspected of causing the problem. The earlier tranquillizers were of the former kind and collectively were known as monoamine oxidase inhibitors (MAOIs). Not only did they have a rather indiscriminate action, they could also interfere with other parts of the body where *monoamine oxidases* were used, and this could lead to unwanted side effects, such as sweating, constipation, weight gain, and a dry mouth.

The alternative approach, of controlling the re-uptake, has mainly targeted serotonin and noradrenaline, and so we have selective serotonin re-uptake inhibitors (SSRIs), noradrenaline re-uptake inhibitors (NARIs), and even drugs that affect both, known as the serotonin-noradrenaline re-uptake inhibitors (SNRIs). There are even some drugs that target specific serotonin receptors, of which fourteen types have been identified. (There are five receptors in various parts of the brain that are activated by dopamine, and at least eight for noradrenaline.)

The most famous of the SSRIs is Prozac and, in treating clinical depression, it has transformed people's lives, so much so that it has been called 'liquid sunshine'. Some patients have reported that it not only cured their depression, but was also life-enhancing. Others have even gone so far as to say that, thanks to Prozac, they felt they had at last discovered their true personalities. So what is this remarkable molecule?

Prozac

Although there have been books and articles attacking this drug, it does offer some remarkable benefits: it lifts depression, it boosts confidence, and it stops the compulsive eating disorder bulimia nervosa.

The Prozac story dates from 1970 when Bryan Molloy and Robert Rathbun of the Eli Lilly Research Laboratories in the United States began a re-search project to find a new antidepressant that did not have the side effects of the early tricyclic drugs. The leader of the research team was Ray Fuller, who was eventually to gain the acclaimed Pharmaceutical Manufacturers Association's Discoverer Award in 1993. Sadly, he died of leukaemia in 1996 after a lifetime of service to the industry, during which time he had published more than 500 research papers in the areas of neuro-pharmacology and neuro-chemistry.

Meanwhile, another Eli Lilly researcher, Hong Kong-born pharmacologist David Wong, was looking at re-uptake mechanisms at synapses, using brain tissue from rats, the object being to find something that would prevent this happening. Wong tried some of the molecules Molloy had made and, on 24 July 1972, he tested the compound coded L110, 140 and found that it was a potent inhibitor of serotonin re-uptake, and a weak inhibitor of norepinephrine re-uptake. Chemical L110,140 was fluoxetine, better known today as Prozac. (Eli Lilly did not immediately pursue Wong's discovery because it was, at the time, more interested in another drug, desipramine, which blocked only norepinephrine re-uptake.)

Tests of fluoxetine undertaken on animals showed it to be safe and effective, and it was then tried on human volunteers in 1976. Strikingly successful, fluoxetine was patented in 1982 (US patent 4,314,081), and eventually put on the market in the United States in 1988, under the trade name Prozac. By 1994, it had become the world's leading antidepressant drug.

Other pharmaceutical companies were soon to market rival SSRI drugs, such as Pfizer's Zoloft (generic name sertraline, US trade name Lustral), to be followed by SmithKline Beecham's Paxil (generic name paroxetine, US and UK trade name Seroxat). Other competitors were Solvay's Luvox (generic name fluvoxamine, US trade name Favertin), and Forest Laboratories' Celaxa (generic name citalopram, US trade name Cipramil).[40] Despite these competitors, Prozac secured 35% of the lucrative and expanding antidepressant market, while Zoloft had 15%, and Paxil 11%.

SmithKline Beecham claimed that Paxil was particularly good at treating people suffering from 'social phobia', in other words people who were acutely shy, and who shunned the company of others because of this. The FDA approved the drug for treating this disorder in May 1999, whereupon the drug company funded a massive campaign to publicize both the condition and its treatment for it. Soon Paxil was being prescribed at levels approaching those of Prozac. Some doctors and psychiatrists were not convinced that there was such a disorder as social phobia,

[40] Prozac is called Prozac in the United Kingdom, the United States, and around the world.

however, and criticized the new drug, saying that some people who are shy by nature don't need treatment because, for many, shyness is a perfectly normal reaction to certain situations. Nevertheless, Paxil was a godsend for those whose shyness constantly interfered with their life, and sales of the drug soared to such an extent that they now exceed $3 billion per year.

Meanwhile, Eli Lilly was taking steps to make Prozac an even better antidepressant. Its activity resided in only one of its two structures known as **enantiomers** (see under 'chirality' in the Glossary), so it would be better to prescribe only the active form alone, known as the R-fluoxetine enantiomer. While this would add to the cost of manufacture, it would have the benefit of reducing the dose required to achieve the same benefit, while reducing the intensity of any side effects. Eli Lilly signed an agreement with the manufacturing company, Sepracor, which holds the patent for making it (US patent 5,708,035).

Prozac is prescribed in doses of 20 mg per day for depression and 60 mg per day for bulimia nervosa. The most famous user of Prozac for the latter condition was Diana, Princess of Wales; it was prescribed to her in 1994. Prozac is not recommended for children, adults suffering from liver failure, or breast-feeding mothers. Some of those who have been given it have suffered side effects such as nausea, diarrhoea, headaches, and so on, but these reactions are rare.

The trouble with Prozac, and the other SSRIs, is that their side effects tend to appear well before the therapeutic benefits, to such an extent that those prescribed these remedies may be inclined to abandon the treatment before they begin to benefit from it. It is still not understood why these antidepressants take so long to work, because they boost the levels of their target neurotransmitters within minutes of being taken. This increase, however, is only the first step in a sequence that eventually leads to relief. Even though the drug boosts the volume of chemical messengers in the synapse gap, these molecules find no target to lock on to because many of the receptors have closed down through lack of stimulation, and it takes time for them to be re-activated.

Prozac gradually builds up in the body until it reaches a certain level, thereafter remaining constant even if the dose is increased, and it continues at this level for several days even if the treatment ceases. Prozac is metabolized by the liver, which converts it to norfluoxetine; this is also an SSRI, acting in the same way as Prozac itself. It is this that partly explains the long duration of the effects of Prozac in the body. The half-life of Prozac is four days, and of norfluoxetine, nine days. This means that, even if no more Prozac is taken, it will be nine days before the level of these actives in the body drops to a half, a further nine days before they fall to a quarter, and possibly as long as a month before they no longer have an effect.

Because Prozac's benefits take so long to kick in, it is necessary to monitor those patients who may be so depressed as to be suicidal during the early part of the treatment but, once Prozac begins to work, the results can be truly remarkable. It is reputed to make the timid socially confident, to make the sensitive person brash, and to turn introverts into extroverts and pessimists into optimists. Not surprisingly, Prozac became one of the most widely prescribed drugs for treating anxiety-related problems, social phobia, panic attacks, agoraphobia, obsessive-compulsive disorder, bulimia, chronic pain conditions, and tension headaches. Prozac gives confidence to people who formerly lacked it, and this type of behaviour was demonstrated scientifically using rats. In a rat colony there is one dominant male; he has twice the serotonin level of other males in the group, and that's something that develops natur-ally. When his underlings also had their serotonin level boosted, they stopped being cowed by him and no longer allowed him to take all the best food.

Within two years of Prozac being available in the United States, doc-tors were issuing more than 60,000 prescriptions per month for it, and, within five years, almost 5 million Americans had taken it. Not surpris-ingly, its benefits were over-inflated, as were those instances when it found its way into the headlines for the wrong reasons—see box. Eventually it ceased to be controversial, and it still continues to deliver remarkable results to the vast majority of its users. Prozac does not have the side effects associated with the older-style antidepressants, some of which were reported to be addictive.

Another benefit of Prozac is that massive overdosing is not likely to lead to death, although some have managed to commit suicide this way after taking between fifty and a hundred tablets; a few of those who even-tually recovered suffered from convulsions. Some who question the use of Prozac say it produces more side effects in users than the older anti-

A murderous attack

Prozac hit the headlines in the 1990s in a way its makers could not have foreseen. It was claimed that it was responsible for the mass slaughter at a printing works in Louisville, Kentucky. The killer was 47-year-old Joe Wesbecker, who had been working at the plant for 17 years, operating one of seven massive printing machines; one of these was so terrible that it was said the foreman demanded oral sex from workers who did not wish to be allocated to it. Wesbecker even had to work double shifts sometimes, and in 1988 he became so depressed that his doctor put him on a course of antidepressants. When these appeared not to work, he tried Prozac in September 1989.

On the seventeenth of that month, Wesbecker, who was on sick leave, went in to work armed with an AK47 and three clips of ammunition, which he used to shoot twenty of the co-workers he most disliked, killing eight of them. He then shot himself with a revolver.

In 1994, the survivors, and the widows of the dead men, brought an action against Eli Lilly, the makers of Prozac. The court found in favour of the company, although the company had already reached an agreement with the plaintiffs to pay undisclosed compensation. Although it seems unlikely that Prozac was in any way to blame for Wesbecker's violent behaviour, the story made the drug notorious for a time.

depressants. This has been answered, however, by the fact that fewer people stop using it because of side effects and that, in any case, the notification of side effects is now a recognized procedure, and patients are more ready to report them.

The side effects blamed on Prozac probably happen because, although it acts mainly on the 5-HT1 receptors, in some people it may also affect the 5-HT2 receptors, causing insomnia and sexual dysfunction, and the 5-HT3 receptors, causing nausea and headaches. Generally, however, Prozac has only a weak affinity for other receptor systems. Some who take Prozac may also develop a rash.

To what extent side effects are due to the drug, or to unrelated causes, can be seen from a series of double-blind tests, in which 1800 patients were given Prozac, and 800 patients were given a placebo. As in such tests, neither the takers of the pills nor those who gave them knew which was which. Of those given Prozac 20% reported nervous headaches, but so did 16% of those given the placebo, and this was the most commonly reported side effect. Of those on Prozac, 18% reported feelings of nausea, compared with 10% on the placebo. Of those on Prozac, 14% reported insomnia, compared with 7% on the placebo. Likewise, 12% reported having diarrhoea, compared with 7%. Only about 5% in each group reported cold-like symptoms (runny nose, etc.). The likelihood of suicide is slightly less among those on Prozac, at 1%, than of those on the placebo (3%), and both are lower than for those given the older style antidepressants (4%).

Perhaps not surprisingly, Prozac's success produced a reaction among those who saw it as part of a 'drug-dependence' culture in which people were becoming reliant on such medication rather than confronting the problems of life. Books have been written that criticize the drug, including Peter Breggin's *Talking Back to Prozac: What Doctors Won't Tell You About Today's Most Controversial Drug*. On the other hand, *Prozac Nation: Young and Depressed in America* is very supportive; it was written by 28-year-old Elizabeth Wurtzel, whose life had been transformed from that of a depressively suicidal drug-taking student to that of a happy and successful journalist and writer.

Peter Kramer's *Listening to Prozac: Antidepressants and the Remaking of the Self* was also very much in support of it, but *Prozac Backlash* by the Cambridge psychiatrist Joseph Glenmullen highlights the side effects of the drug, claiming that it produces facial and body tics and sexual dysfunction in over half of those who take it. Moreover, he says that when users stop taking it they can suffer withdrawal symptoms such as dizziness and anxiety.

Those who prefer complementary medicine, and deny themselves the use of pharmaceutical drugs such as Prozac, may still seek the kind of relief they bring. There are natural alternatives to taking Prozac, and these take several forms.

If there is a shortage of messenger molecules in the brain, then we might try to correct this by taking more of the amino acids tyrosine and tryptophan as dietary supplements, because the former is the molecule from which dopamine and noradrenaline are metabolized, and the latter is the precursor to serotonin. Tyrosine is plentiful as a component of the protein in dairy foods, eggs, salmon, nuts, and processed meats, and it even occurs as the free amine in oranges, plums, and tomatoes. Tryptophan is also at relatively high levels in the protein of grains, nuts, meat, beans, fish, and dairy products, and again, there are some foods such as cheese and sausages that may contain the free amine. A normal diet will provide all the tyrosine and tryptophan that the body needs to make its chemical messengers, but it is just possible that a depressed person on an unbalanced diet might well respond to an increase in these types of protein.

If dietary changes fail to relieve your depression, then you can try herbal remedies that can be bought over the counter in a health food store, and these may be effective. We know that some herbal remedies really do contain active ingredients, and operate in the same way as the pharmaceutical drugs. Two of the most popular natural remedies are St John's wort and kava.

St John's wort (*Hypericum perforatum*) is a native plant of temperate regions, and it even grows on the sides of mountains in tropical regions; it has oval to oblong leaves and bright-yellow flowers. In medieval Europe its petals were infused with boiling water, and the concoction drunk to restore one's spirits. It is widely used in Germany as a mild antidepressant, and with medical approval because it contains the molecule hypericin, which has been shown to inhibit *monoamine oxidase*. Like the antidepressants mentioned above, however, it takes time to work.

Provided you are not taking any other form of medication, you might well find St John's wort beneficial, and it can be taken as a tea, as a tincture, or in capsule form. If you are taking a prescription drug, however, then you should be aware that there are risks. Research by the US National Institutes of Health has shown that St John's wort will interfere in a negative manner with some of them. The drugs likely to be affected are birth control pills, cholesterol-lowering medicaments, and drugs that prevent organ transplant rejection. Nor is St John's wort without its side effects: indigestion, allergic reaction, and tiredness, and it can make its users more sensitive to sunburn.

Kava (also known as kava kava; *Piper methysticum*) is a leafy green plant which originated in the islands of the South Pacific, and which was used by natives there in the form of an infusion made from the ground-up roots of the plant. Today, it is generally available in the form of capsules and is popular as an antidepressant. There are claims that double-blind tests, using doses of 400 mg, show it to be as effective as Valium. The active constituents of kava are a collection of ring compounds known as lactones and pyrones, although the amounts of these present can be very variable. They do have pharmacological effects, notably acting as sedatives, muscle relaxants, and painkillers, as well as having a calming effect on mood. The natives of the Pacific islands continue to drink kava kava on social occasions, although in parts of the region there are laws governing its sale and use, as with nicotine and alcohol. In particular, kava should not be used by pregnant and breast-feeding women, nor by those who drive or operate machinery. More serious is the advice of the US Food and Drug Administration, which warns that users of kava can suffer severe liver damage, although this is rare. The less serious side effects of kava are nausea, vomiting, loss of appetite, and abdominal pains, although again not all those who take kava suffer from these.

If St John's wort or kava are not to your taste, there are other treatments for mild depression that you might try before resorting to your doctor. Homeopathy, acupuncture, Chinese medicine, reflexology, or aromatherapy may well give you the support you need to get through a difficult phase of life, but there are some people for whom such panaceas would be of no use; nor are Prozac-type antidepressants much use either. For them only a more powerful medical treatment can hope to succeed, and that's when one of the most curious remedies of all is the best solution: lithium.

A little light-headed: lithium therapy

The lightest of all metals can have a profound effect on the brain, to the extent of enabling those with the most disabling type of depression to lead a normal life.

It has been estimated that, in the United States and Europe, close to half a million people take lithium to alleviate the mood swings of bipolar (manic) depression. This metal can play a biological role in controlling neurotransmitters in the brain and so have a calming effect on mental disorders that are too overwhelming to respond to the usual anti-depressants.

Lithium therapy is not given to correct a deficiency of an essential element. Indeed, the human body appears to have no biological need for this metal, but we do have measurable amounts within us. Lithium is widely dispersed in Nature and we take some in each day with our food and drink. Typical values in food are, for example, 0.5 p.p.m. (dry weight) in corn and cabbage, 0.3 p.p.m. in lettuce, and 0.2 p.p.m. in oranges, and potatoes can have up to 30 p.p.m. if they are grown in soil where there is a lot of lithium. We absorb a little lithium from the food we ingest, but that which is absorbed is also excreted easily, so it does not accumulate in our tissues or organs. The average adult has around 7 mg of lithium in his or her body, and the level in the blood is around 4 p.p.b., which shows how relatively unimportant it is. The low level is not surprising considering how little there is in the food we eat.

We cannot totally dismiss lithium as having no biological role in animals, and at least one experiment in the past has suggested it may have a part to play. Goats fed a lithium-free diet from birth were found to put on less weight than those that were given a normal diet.

Though unrecognized as such, lithium has a long pedigree in treating disease. The waters of ancient Ephesus, then a Greek city, now part of Turkey, were reputed to be good for the brain—or so wrote Soranus of Ephesus, who was born there in the second century A D and practised as a doctor in Rome. He specialized in the diseases of women, and his book *On Acute and Chronic Diseases* has a section on the treatment of nervous disorders. In fact, the water around Ephesus has a higher than normal amount of lithium dissolved in it, but not to a level that could have had much effect on those who drank it.

The lithium story really began towards the end of the eighteenth century. The first lithium-containing mineral to be discovered was petalite, which was found on the Swedish island of Utö by the Brazilian scientist Jozé Bonifácio de Andralda e Silva, on a visit there in the 1790s. When it was analysed for the elements it contained, these totalled less than 100%, which naturally puzzled its investigators. The problem was solved at Stockholm in 1817 when Johan August Arfvedson (1792–1841) agreed to analyse it more closely. He spent many days doing this and could still account for only 96% of its weight, but he made the obvious deduction from his research. In January 1818 he explained the unaccounted-for 4% as a new metal and, announced the discovery of lithium, which he said was a new alkali metal—and he was correct. He named it lithium because it had been found in a 'stone', the ancient Greek word for which was *lithos*.

That same year, it was reported that when sprinkled into a flame, lithium gave off a beautiful red colour, and this proved to be a useful and quick test for detecting its presence. Soon after its discovery, lithium was found in other minerals, and even in spa waters at Karlsbad, Marienbad, and Vichy. To measure minute traces of lithium required a more sensitive technique than the flame test, and this was eventually provided by the spectroscope, which allowed lithium to be revealed by the characteristic red line in its spectrum. This resulted in the metal being detected in seawater in 1859 (in which its concentration is only 0.17 p.p.m.), and in subsequent years it was also shown to be present in grapes, seaweed, tobacco, various vegetables, milk, blood, and human urine.

Although lithium is an alkali metal, like sodium and potassium, Arfvedson was not able to separate it by electrolysis, as Humphry Davy had done for the other alkali metals. William Brande was able to do that in 1821, however, but obtained such a tiny amount of the metal that he had not enough to record its properties. That had to wait until 1855, when the famous German chemist Robert Bunsen (1811–99), and the lesser-known British chemist Augustus Matthiessen (1831–70),[41] succeeded independently in isolating enough of the metal to study it, and they achieved this by the electrolysis of molten lithium chloride.

Lithium metal is not used as such because it is quickly attacked by moisture. Rather surprisingly, and unlike sodium, it does not react with oxygen in the atmosphere unless heated to 100 °C, but again unlike sodium, it does react with atmospheric nitrogen to form lithium nitride.

Being the lightest of all metals, lithium produces alloys that are also light, the most important one being that with aluminium, which can be used to reduce the weight of aircraft and space vehicles, and thereby save fuel. Lithium is also used to make long-life batteries for electronic equip-

---

[41] London-born Augustus Matthiessen studied in Germany, and eventually became Professor of Chemistry at St Mary's Hospital in London in 1862 at the early age of only thirty-one. He later moved to St Bartholomew's Hospital, London, but was there only a year before he died, aged thirty-nine.

ment where compactness and lightness are all-important. Such batteries generally have lithium as the anode and iodine as the solid electrolyte, and can have a lifespan of ten years.

Lithium carbonate (Li_2CO_3) is the single most important compound of lithium and, being of low solubility in water, it is easy to prepare and separate. It is used in aluminium refining, in glass, enamel, and ceramics, and it is the starting material from which other lithium compounds are made. Purified lithium carbonate is the form in which it is prescribed as a treatment for extreme forms of depression, and, although this is only a minor use of lithium in terms of quantity, it is arguably the most beneficial use of all. An intriguing aspect is that we still do not entirely understand how it works its magic in the brain.

In the nineteenth century, lithium enjoyed a vogue as the treatment for gout, the painful condition in which sharp crystals of uric acid form between the joints, especially those of the feet. Uric acid is not very soluble, so that once the crystals form they may take a long time to dissolve again. Because the lithium salt of uric acid is very soluble, however, it was reasoned that drinking lithium-rich waters should assist its removal from the body—or so said a Dr Ure in 1843. This suggestion was popularized by the eminent Victorian physician Sir Alfred Garrod (1819–1907), who had observed just how quickly a solution of lithium carbonate dissolved uric acid, at least in the test-tube, and he recommended taking large doses. The theory that lithium was the best way to treat gout held sway for more than half a century, until it was eventually proved to be of no benefit whatsoever. Indeed, in 1912 a Dr Pfeiffer showed that taking lithium compounds actually slowed up elimination of uric acid from patients with gout.

Meanwhile, other doctors were trying lithium for other conditions. In 1864, lithium salts appeared in the *British Pharmacopoeia*, and these included lithium carbonate, which could be used to treat indigestion, and lithium citrate, which acted as a diuretic. In 1871 a William Hammond recommended large doses of lithium bromide for 'acute mania' and 'acute melancholia', and a Danish neurologist, Carl Lange, prescribed a mixture of alkaline salts, of which lithium was a major component, saying it worked well for 'periodic depression'. These suggestions were well ahead of their time and were not taken up by the medical world. Then in 1949 an Australian doctor, John Cade, accidentally made the breakthrough that was to see lithium become the best treatment for such conditions.

Cade was experimenting with guinea-pigs, injecting them with urine taken from manic-depressive patients in the hospital where he worked, in the hope of proving that their condition was actually caused by excess uric acid. The animals were also given injections of the lithium salt of uric acid. Cade found to his surprise that the guinea-pigs, normally highly strung animals, became so calm that they could be turned on their backs and would lie placidly for hours at a time.

Cade then tried lithium carbonate on his most troublesome patient, who had been admitted to a secure ward five years earlier. He found that the man responded so well that within days he was transferred to a normal hospital ward, and within two months was able to return home and take up his old job. When Cade's findings were published, other doctors began treating manic-depressive patients with lithium carbonate and found it cured them as well. By 1960, the treatment was accepted throughout Europe, and by 1970 in the United States as well, despite the earlier adverse publicity there about lithium chloride (LiCl) causing the deaths of those using it as a salt substitute.

After World War II, LiCl had been marketed as Westsal for use by those with heart disease who were on salt-free diets. In 1949 several patients died of lithium poisoning, and urgent warnings about Westsal were issued in the press and over the radio. The FDA even banned the use of lithium salts altogether, and carried out its own research on them, showing that in large doses lithium can adversely affect the kidneys. This ban persisted for about 20 years, until it had to be lifted if the benefits of lithium therapy were to be enjoyed by US doctors and their patients.

Lithium is generally prescribed as lithium carbonate in 250-mg tablet form, and it is most effective when the level of lithium in the blood is between 0.6 and 1.2 **millimoles** per litre.[42] (See 'units' in the Glossary.) It is dangerous to exceed a dose of 2.5 grams at any one time, and a blood level of 15 mg per litre indicates mild lithium poisoning. There may be long-term effects of taking lithium because it can result in permanent damage to the kidneys, which is why a person is generally only pre-

---

[42] This is equivalent to between 4.2 and 8.4 mg per litre, which is the same as saying between 4 and 8 p.p.m.

scribed it for a maximum of five years, and why he or she needs to be regularly monitored throughout this period.

It is rare these days for doctors to prescribe a medicament whose mechanism of action in the body is not understood, but lithium comes into that category although there are theories as to how it works. The first theory, and still regarded as the most likely explanation, came in the mid-1980s from Michael Berridge and co-workers at the Unit of Insect Neurophysiology and Pharmacology in Cambridge, England. They said that lithium depressed the level of inositol, a sugar-like molecule that is found in cell membranes throughout the body, including those in the brain. Other parts of the body can get inositol as inositol phosphate from the food we eat, especially from cereals, vegetables, and citrus fruits, but none of this inositol phosphate reaches the brain, which therefore has to make its own. Its production and recycling in that organ is susceptible to lithium interference because it can penetrate into brain cells.

Inositol phosphate is responsible for passing signals from the outside to the inside of a cell, which it does with the help of calcium ions. After the inositol phosphate has done its job, it is recycled back to inositol by the enzyme *inositol monophosphatase*. It is this enzyme that really controls the sensitivity of a cell and the brain to external stimuli, and if it is too sensitive then the result is mood swings and manic behaviour. Lithium dampens down the activity of *inositol monophosphatase*, but only in those nerve cells where there is too much, which explains why lithium has no effect on those who are not depressed, and why taking it won't give you a lift.

An additional explanation as to how lithium works was suggested by a team of researchers at Merck Sharp & Dohme in Harlow, England. Howard Broughton, Scott Pollack, and John Atack examined the way lithium affects the behaviour of the enzyme and concluded that it did so by occupying a site in the molecule that would normally be occupied by magnesium, once this happens, the enzyme ceases to function as it should.

At first sight this interchangeability seems at odds with the chemistry of these elements, because lithium and magnesium are in adjacent groups of the periodic table. Lithium is in group 1 and magnesium in group 2, giving differently charged ions, Li^+ and Mg^{2+}. Nevertheless, they do have chemical similarities linked by what is referred to as a **diagonal**

relationship between elements in different groups (see Glossary). It is therefore possible that the environment in an enzyme that would attract and hold a magnesium ion would also be equally attractive to a lithium ion, but once that was present the activity of the enzyme would be rendered useless until the lithium departed. Replacing magnesium with lithium will depend on the concentration of lithium in the body, and this would need to be kept at a high enough level so that it could compete successfully with magnesium and deactivate enough overactive nerve cells in the brain, and so calm the patient.

There are other theories as to how lithium operates. Dr Adrian Harwood of University College London believes that it disables the enzyme *glycogen synthase kinase-3*, which is part of a signal pathway.

Clearly, there is still work to be done in understanding how lithium acts in the body; until that is known, there is no possibility of finding a safer alternative medication, or of making lithium treatment more targeted, so that less need be prescribed and the treatment can be continued for as long as needed. However lithium operates, it appears as yet unchallenged in the treatment of extreme forms of depression.

In 1991, in the book *Lithium in Biology and Medicine*, it was noted that epidemiological studies in the United States indicated that areas where the drinking water had relatively high levels of lithium, that is, in excess of 70 p.p.b., were areas with lower levels of crime and fewer suicides. The obvious deduction was that authorities might adjust the level of this metal in the water supply where crime was high and suicides more frequent. This use of lithium will never happen, if only because this metal is never going to be cheap enough, even if you might persuade a water company to carry out such an experiment for the many years needed to prove it had an effect.

Lithium may also be involved in other medical treatments; for example, lithium-based creams have been shown to inhibit genital herpes, thereby alleviating a condition that might well lead its victims to need other antidepressant treatments.

Alzheimer's disease and aluminium

What causes Alzheimer's disease? More than twenty risk factors have been suggested, mainly on the basis of epidemiological evidence, but for many years the finger of blame was pointed at aluminium—wrongly, as it turned out.

We know that the incidence of Alzheimer's disease (AD) is definitely linked to old age and that some people are more at risk by virtue of a genetic predisposition. AD may possibly be made worse by vitamin deficiency and lack of mental stimulation, but these are not the *cause* of the disease. The real cause is as yet unknown, but there are now strong contenders: enzyme malfunction, genetic fault, microbial invasion, or toxins such as metals. It was this last category that seemed at one time to explain the disease and to offer a way of preventing it. The culprit was alumin-ium. For 25 years, a lot of time, money, effort, and publicity was directed at controlling the impact of this metal on our lives, but it was 25 years of wasted endeavour.

The first person to be diagnosed with AD was Auguste D,[43] a 51-year-old German woman who in 1901 became forgetful, disoriented, had difficulty in reading, and who developed symptoms of obsessive jealousy over her husband. She was admitted to the Frankfurt Asylum, where her case was referred to the neurologist Alois Alzheimer (1864–1915). He documented her continued decline, noting that she was given to outbursts of screaming and foot stamping, was unable to remember the names of her family, and that she eventually became totally apathetic and incontinent.

After her death, in April 1906, her brain was sent to Alzheimer, who was now at the Royal Psychiatric Clinic, part of the Munich Medical School. He found it to be greatly shrunken, which surprised him because she was a relatively young woman. He examined samples of its tissue, and by staining them with silver and inspecting them under the microscope he discovered curious deposits in the damaged nerve cells. Alzheimer reported his finding at the 37th Conference of German

---

[43] Her surname was not revealed.

Psychiatrists in Tübingen in 1907, and published them in the leading German psychiatric journal, calling his paper 'A new disease of the cortex'. The disease was named after him in 1910 by the eminent German psychiatrist Emil Kraepelin, who referred to it as Alzheimer's disease in the eighth edition of his influential *Handbook of Psychiatry*, and the name stuck. Neither the symptoms, nor Alzheimer's findings, were new, however, but it suited Kraepelin to publicize them as a new disease because this gained the Munich Medical School valuable publicity and so ensured a continued flow of funding into the institution.[44]

Today we know much more about AD. There are two types referred to as the familial and the sporadic, in other words, the inherited and the non-inherited forms. Both follow the same relentless progress and with the same symptoms. AD is characterized by two types of abnormal formations within the brain: fibrous tangles within brain cells and spherical senile plaques in the spaces between the cells. The former are composed of an abnormal protein called *tau*, while the latter are clumps of another protein called amyloid.

Tau protein is produced mainly by nerve cells in the brain, to help it to maintain the microtubules that act as conduits along which molecules can be rapidly transported down the nerve cell's axons. But when *tau* protein is overloaded with phosphate it clumps together to form a tangled mass, though this is observable only with the aid of a powerful electron microscope.

Amyloid fibrils are usually blamed for killing neurons in the brain, but they may be more of a symptom than a cause of AD. At least that's the theory proposed by a group of researchers headed by Peter Lansbury Jr at Harvard Medical School in the United States. He claims that there is a soluble precursor, a shorter chain of protein, that does the damage by creating holes in cellular membranes, and that these holes allow access to calcium and other ions to such an extent that they may then enter and kill the neuron cell.

Most researchers into AD, however, still believe that the plaques are more dangerous, and that it is these that lead to neuron malfunction and ultimately death. In 1984, the plaques were found to consist of beta-amyloid peptide, a protein waste product that the brain finds difficult to

---

[44] Some sceptics believe that this misuse of the media to promote funding for research continues even today, and not only in Germany.

eliminate, and when this happens it builds up as deposits. A protein in the plasma binds to these deposits and protects them from being broken down for recycling by protein-digesting enzymes. This protecting protein is known as serum amyloid protein and is produced by the body at the rate of 50 to 100 mg a day, with any excess being disposed of by the liver. It protects the unwanted amyloid deposits just as well as it protects other amyloid proteins. As we shall see, removal of serum amyloid protein from the bloodstream offers a possible way of getting rid of the plaques.

The parts of the brain most affected in AD are the outer cortical regions, and the hippocampus and the amygdala, which are deep within the brain. The loss of nerve cells in the hippocampus region cannot be replaced, and this loss leads to the most obvious symptom acute memory loss, for which there can be no cure. The best we can hope for is to prevent it or to arrest it once it has been diagnosed.

AD now afflicts more and more people because it is a disease of old age, and in developed societies there is invariably an ageing population. In the United States there are 4 million sufferers, in the United Kingdom around half a million, and the global total is now in excess of 12 million. The incidence of AD rises steeply with age and as many as 15% of people over eighty are afflicted. The condition is rare in those aged sixty to sixty-five, when only 0.1% of people are affected. (The genetic form of the disease accounts for 5% of AD cases but that can even start in people as young as thirty-five.)

The environmental agent that was for a long time regarded as the most likely cause of AD was aluminium. What fingered this metal was its role in a disease with similar symptoms: dialysis dementia. As its name implies, dialysis dementia afflicted patients who were given kidney dialysis treatment, in which they were connected up to a machine that was able to flush toxins from the blood, thereby carrying out the function their kidneys were no longer able to do.

In the 1960s, it was noticed that those who were given dialysis treatment two or more times a week began to suffer from dementia; in other words, they began to behave oddly, talk illogically, became forgetful and confused, and were clumsy. The cause of dialysis dementia was aluminium, which was being dissolved from the dialysis equipment and going directly into the bloodstream, and thence to the brain. At about this time, a group in Newcastle, England, reported that plaques taken from the brains of such dementia victims contained high levels of aluminium. Moreover, when aluminium salts were injected into rabbits and cats, tangles of fibrous material were produced in their brains.

The aluminium that caused dialysis dementia did not result in the types of fibrous material or amyloid plaques associated with AD, however, and in the rabbits and cats the fibrils were present in other parts of the nervous system as well as the brain. Indeed, the effects were more like a toxic response to the metal, and the same was true of humans. When those suffering from dialysis dementia were treated with drugs to remove aluminium from the body they recovered. Dialysis dementia was quite unlike AD in this respect, but when it was reported that aluminium had also been found in the plaques taken from the brains of AD victims, it seemed logical to suppose that this metal might also be the cause of that disease. The link between aluminium and AD was widely publicized, and widely believed. The bandwagon began to roll.

Not everyone subscribed to the theory. Sir Martin Roth, Emeritus Professor of Psychiatry at Cambridge University and a recognized expert on AD, did not support it and publicly declared as much, but his was a lone voice. For those who did believe it, and there were many, it was clearly going to be an uphill struggle to remove the impact of aluminium, because it was now part of everyday life and a metal about which the public had fond memories, at least in Britain.

Aluminium is one of the lightest of metals, with a density of only 2.7 grams per cubic centimetre, which is almost a third that of iron. Little wonder, then, that this metal was used in aircraft manufacture. Indeed, the aluminium Supermarine Spitfire fighter that defeated the Nazi *Luftwaffe* over the skies of Britain in 1940 and 1941, and thwarting Hitler's planned invasion of the United Kingdom, was a fitting tribute to what aluminium could achieve.

Aluminium is the third most abundant element in the Earth's crust, coming after oxygen and silicon, and it is the most abundant metal. Because its oxide is so insoluble, only a little of this metal enters natural waters. A small amount of aluminium gets washed into the rivers and thence to the seas each year, but the concentration in seawater is only about 0.5 p.p.b. because it precipitates to the bottom. If soil becomes acidic and the pH drops below 4.5, then aluminium becomes more soluble and this can affect plants and crops by reducing root growth and phosphate uptake. This was the reason why acid rain was thought to be so damaging to plants.

Some aluminium gets into the food chain, but most passes through our gut without being absorbed because it is there mainly as insoluble material. There are some soluble salts of aluminium, such as the citrate, so there is the possibility of some aluminium being able to pass into the bloodstream if it meets citric acid from citrus fruits, but this serves no useful purpose. Aluminium has no role in human metabolism, but given our daily intake it is perhaps also not surprising that the average person contains around 60 mg.

In everyday life we generally encounter aluminium in the form of cans or foil, and it is easily recycled. Its compounds are also useful, and especially alum (which is potassium aluminium sulfate), aluminum sulfate, aluminium hydroxide, and aluminium oxide. The first of these has been used for centuries as a mordant in dyeing and to staunch bleeding; the second is used in water purification; the third you might take to relieve indigestion; and the fourth you might wear as jewellery. (Several gemstones are made from the clear crystal form of aluminium oxide. The presence of traces of other metals creates various colours: iron produces the yellow of topaz, cobalt creates the blue of sapphires, and chromium makes rubies red. All these are now easy and cheap to manufacture artificially.)

All in all, we have been using aluminium for centuries—see box on page 172—but could we inadvertently have been putting our mental health at risk by doing so?

Because aluminium is so abundant in soil, all plants absorb a little of it. Tea bushes absorb a lot and alum is used as a fertilizer in tea plantations. Tea provides a significant amount of the aluminium in the diet of those who prefer this beverage. Plant foods with most aluminium are spinach (104 p.p.m., dry weight), oats (82 p.p.m.), lettuce (73 p.p.m.), onions (63 p.p.m.), and potato (45 p.p.m.). Processed foods have even more, especially processed cheese, which has 700 p.p.m., while sponge cakes and biscuits have a lot because sodium aluminium phosphate is used as a leavening agent in their baking.

At the time the aluminium scare started, there were several ways in which food came into contact with this metal: aluminium cooking pans,

Aluminium in history

As long ago as the first century AD, the Roman army doctor Dioscorides (c. 40–c. 90) wrote the medical book *De materia medica*, in which he recommended alum to stop bleeding, and to treat various skin conditions such as eczema, ulcers, and dandruff—although in these cases it would be of little benefit. The ancient world got its alum from naturally occurring deposits in Greece and Turkey. For centuries it was traded and exported via Constantinople, but in the Middle Ages it was discovered that it could also be made from clay and sulfuric acid. The discovery of alunite, a potassium aluminium sulfate rock, at Tolfa in territories controlled by the pope, led to a papal monopoly of its manufacture in Europe from the 1460s onwards. The industry employed 8000 men, produced 1500 tonnes of alum a year, and generated a large revenue for Rome.

This state of affairs continued until English alum production started in the early seventeenth century, in north Yorkshire where a large deposit of alum shale was discovered. Production there broke the papal monopoly and the price of alum fell dramatically. Production continued in Yorkshire for more than 250 years. Despite alum-making being the first true chemical industry, there was little understanding of the nature of the product or the process of its manufacture.

In the eighteenth and nineteenth centuries alum was used by paper-makers as a preservative, by doctors to stop bleeding, by scientists to preserve anatomical preparations, and by dyers as a mordant to fix dyes to fabrics (the aluminium ions stick to the fibres of the cloth and the dye sticks to the aluminium ions).

foil, food containers, take-away trays, and drinks cans. Rhubarb cooked in aluminium pans was able to dissolve some of the metal by virtue of the oxalic acid it contains. Indeed, this was the traditional way to clean such pans when they developed a dark-coloured stain through long usage.

Given all these sources of the metal, perhaps it is not surprising that the average daily intake of aluminium is between 5 and 10 mg, but only a small fraction of this is absorbed, probably less than 0.01%. Dietary components such as silicate, fluoride, and phosphate prevent its absorption by forming insoluble salts. Even that which is soluble finds it almost impossible to pass through the gut wall into the bloodstream, where levels of aluminium are typically measured only in fractions of p.p.m. In any case, our kidneys are very good at removing the metal. Consequently, the body's metabolism has little need to defend itself against a massive insult from this metal, because under normal conditions this would never occur. In dialysis patients it did occur: aluminium attached itself to a molecule in the blood, called transferrin, and in this way it gained entry to the brain.

Despite the historical evidence that people were not at risk from aluminium, a campaign of almost religious fervour developed to root out all possible sources of contamination. Aluminium was to be found in some quite surprising places, including food additives. In Europe there were four permitted ones, and these had the code numbers E173, E541, E554, and E556. The first of these is aluminium metal powder itself, which was used to make 'silver' cake decorations and to cover sugared almonds. The second is sodium aluminium phosphate, which is used as an emulsifying agent. E554 is aluminium sodium silicate and E556 is aluminium calcium silicate, both of which may be added to powdered foods, such as coffee whiteners, cake mixes, and packet soups, to prevent them caking into a solid mass.

Less well publicized was the intake of aluminium from certain medical treatments, namely indigestion mixtures and medicines prescribed for gastric and duodenal ulcers, which were based on aluminum hydroxide. If the aluminium hypothesis was correct then it might have been expected that people on such medication would show a greatly increased risk of AD, but that appears not to have been the case. (Two studies appeared to show a slight risk but another study showed no risk at all.) Aluminium hydroxide was supplied as a colloidal suspension or in tablet form. The liquid form contained around 4% and was flavoured with peppermint oil and saccharin, and up to 100 ml a day (in doses of 15 ml every two to four hours) was commonly given, representing a daily intake of 4000 mg. In tablet form it was given as 500 mg tablets.

Even less well publicized was aluminium's use in water purification where, as a solution of aluminium sulfate, it can purify drinking water, as well as clean up waste waters and remove phosphate from sewage. It was the first of these uses that suddenly put aluminium sulfate in the media spotlight in the United Kingdom, when the driver of a tanker loaded with it arrived at a local water company, where there was no one on duty to tell him what to do. The result was what came to be known as the Camelford incident.

For more than a century, aluminium sulfate had been part of the water industry, used on a large scale as a flocculating and clarifying agent. It works by forming a voluminous insoluble precipitate of aluminium hydroxide and calcium sulfate, known as the 'floc', which traps dirt and bacteria as it settles out of solution, leaving the water crystal clear. It leaves behind 0.2 mg of dissolved aluminium per litre, however; this is only 0.2 p.p.m., although this value can be expressed as the seemingly more alarming amount of 200 p.p.b. (The European Union and the World Health Organization recommend this as a maximum level in drinking water.) Water contributes about 0.03 mg to the total of 5 mg or so of aluminium that the average person consumes each day in their diet; that is, less than 1%.

Aluminium sulfate is prepared by reacting aluminium hydroxide with sulfuric acid, and is generally supplied to water-works as a concentrated solution that is stored in underground tanks and used as required. It is first diluted, and then added to the water to be treated, whereupon it reacts with the calcium bicarbonate that is naturally present in the water to form the 'floc'.

In July 1988, at Camelford in north Cornwall, England, our tanker driver arrived at the local water-works with a 20-tonne load of concentrated aluminium sulfate solution, and not being familiar with the site he pumped it into what he thought was the storage facility. In fact, he pumped it directly into the water mains. In so doing he contaminated the water supply that served 20,000 people. The affected water contained up to 50 p.p.m. of aluminium, although at times the concentration was ten times this, and some residents actually drank tea and coffee made with it. Some of those affected have campaigned ever since for compensation, citing a variety of ill effects that they say it caused. To date, various commissions of investigation have found no evidence to support the campaigners' contentions. Indeed, a paper from the Childhood Cancer Research Group of Paediatrics at the University of Oxford, published in the *British Medical Journal* in May 2002, showed that in the ten years following the incident the death rate among Camelford residents exposed to the contaminated water was slightly lower than that of neighbouring areas where people had not been so exposed.

The presence of aluminium in ordinary drinking water came as a revelation to many people and became a cause for concern. Epidemiological surveys soon claimed to find links between it and the incidence of AD (nine out of thirteen published studies came to this conclusion), despite the difficulty of assessing the aluminium coming from other sources in the diet. Nevertheless, the anti-aluminium campaigners were now demanding an end to its use in water treatment, and companies began to comply with their demands.

The findings were also at odds, however, with the widespread use of aluminium hydroxide as a prescribed treatment for peptic ulcers and as an over-the-counter treatment for acid indigestion, both of which involved taking up to 3 *grams* per day, an amount that had up to then been regarded as perfectly safe by doctors. Some patients suffering from hyperphosphataemia (too much phosphate in the blood) were being prescribed even higher doses of aluminium hydroxide because this reacted chemically with any phosphate in their food to form insoluble aluminium phosphate, thereby preventing its being absorbed across the gut wall.

Over-the-counter indigestion remedies were available in the form of Aludrox liquid and Actal tablets. These were sold as treatments for an upset stomach, and, mixed with magnesium trisilicate, they also counteracted heartburn and prevented constipation.

Nevertheless, other uses of aluminium were being identified with a view to ending them. Aluminium in the form of alum was also to be found in bathroom cabinets. This compound acts as a powerful astringent; that is, it staunches bleeding by precipitating proteins, and a styptic pencil is often used on cuts caused when shaving. A weak solution of alum can be used to harden the skin and prevent sore feet. Basic aluminium chloride acts as a good antiperspirant, and was preferred in cosmetics because it is less irritating to the skin than alum itself.

By 1990, the campaign against aluminium was building to a crescendo when suddenly it was undermined: more sophisticated chemical analysis of brain plaques taken from AD victims showed them to contain none of the metal at all! This research was done at Oxford in 1992 and repeated in Singapore in 1999, in both cases using nuclear microscopy. This technique is particularly good at detecting the presence of a metal like aluminium, but no significant amount of aluminium could be found in the brain tissue. The new work was published by Frank Watt and co-workers in the highly regarded science journal *Nature*, and later backed up by further research.

Watt concluded: 'the results, taken in conjunction with previous work on the analysis of neuritic plaques using nuclear microscopy, suggest that there are no grounds for considering aluminium as a factor in the aetiology of Alzheimer's disease. It is possible that earlier investigations utilising treated tissue, in which aluminium has been detected in neuritic plaques and neurofibrillary tangles, may have been due to contamination or elemental redistribution.'

More work needed to be done before aluminium was completely in the clear, and in the year 2000 a group of volunteers was given large daily doses of aluminium hydroxide for 40 days, during which time their urine was monitored for aluminium. This showed that they were excreting absorbed metal at more than ten times the normal rate, and in some cases at twenty times the rate, but that this was having no effect at all on the body's immune system. The obvious conclusion was that aluminium is well tolerated by the body, which is perhaps not surprising considering that this element is so abundant in the natural environment.

For most people in the United Kingdom the best reassurance that aluminium was no longer a threat came in 1998, when the popular television cook Delia Smith recommended aluminium frying pans and saucepans in her programmes, thereby sending sales of these soaring. Behind the scenes, water companies were reinstating aluminium sulfate, and once again it continues to be the preferred treatment for water purification. Even those water companies that had changed to using iron sulfate in place of aluminium sulfate have now returned to their former method of purification.

So, is aluminium safe? Not entirely, and those with kidney malfunctions should aim to avoid it. Even though 99% of ingested aluminium passes through the body unabsorbed, they, unlike most people, cannot excrete the 1% that is absorbed. The rest of us need no longer fear aluminium, although as we get older we may fear the onset of AD. Is there any way we might avoid this condition by changing what we do? Probably not. The best we can hope for is that medical science will come up with a cure, or better still, something to prevent its onset. If that is to happen, then we need to understand fully the causes of the disease. There are already several clues as to what these may be.

Treatments for Alzheimer's disease (AD)

Alzheimer's disease is an affliction that many dread, and that even more people have to live with in terms of its affecting a loved one; its onset is difficult to diagnose and its progress almost impossible to prevent or delay. But there are new drugs coming on to the market that may well change all that.

As we have already noted, the most obvious physical change in the brains of those with AD is the formation of rogue proteins, and these are deviants in that they have 'folded' incorrectly; folding is the technical term for the way in which a protein chain arranges itself into the structure required to carry out its functions within the body. Proteins are manufactured by enzymes all the time, and for lots of purposes they do this by linking together amino acids into long chains. It is not enough just to bring the right amino acids together, however; the chains have then to 'fold' correctly. When proteins fail to fold to their correct structure, things really start to go wrong. The mis-folded proteins appear as masses of tangled threads, and they can be deposited in all kinds of body tissue. They are more properly known as amyloid fibrils.

Evolutionary selection produced ways of folding proteins into useful structures, despite their natural tendency to want to fold into fibrils. When our bodies lose control over protein folding, fibrils form and the consequences manifest themselves as Creutzfeldt–Jakob disease (the human form of BSE, or mad cow disease), AD, and Parkinson's disease. Some victims of these conditions can end up with *kilograms* of mis-folded protein fibres in their bodies. Clearly a drug that stopped this happening would seem to offer most hope of preventing the disease, and already there are signs of ones that might be able to do this.

Meanwhile, there are other ways in which AD can be alleviated. Anti-depressants can counter one of the early symptoms of the disease, chronic depression, as victims begin to realize that they are losing control of their mind. Non-steroidal anti-inflammatory drugs (NSAIDs), such as ibuprofen, may help protect against AD, because inflamed brain cells tend to collect around plaques and they then harm healthy neurons.

Many NSAIDs can be bought over the counter as proprietary drugs, and are commonly used to treat headaches (e.g. paracetamol), muscular pains (ibuprofen), or both (aspirin). A natural NSAID is curcumin, which is derived from the curry spice turmeric.

Reactive oxygen metabolites (ROMs) can damage and kill cells, and as we age our body finds it harder to remove these natural chemicals. Hearing loss in old age is due to ROMs, and the degree of hearing loss gives some indication of the extent to which the body is coping with their elimination. There is some evidence that antioxidant vitamins are beneficial, and if they are in short supply in the diet, this might well exacerbate AD.

AD affects those parts of the brain rich in the chemical messenger acetylcholine (ACh), which is produced by an enzyme called *choline-acetyltransferase*. When ACh has done its job it has to be eliminated, and this is done by another enzyme, *acetylcholine esterase*. The danger is that, as our brain ages, this enzyme remains effective at removing ACh but *cholineacetyltransferase* slowly loses its ability to produce it. For most people this is not a problem because we have the capacity to produce more ACh than we actually need, but for some people the rate of loss reaches a crisis point and then there is just not enough ACh to do all that is required. A link between reduced levels of ACh and the onset of AD was made almost thirty years ago, in 1976.

An obvious target for drug research has been to find chemicals that are capable of blocking the ACh-removing enzyme, thereby preserving the amount of ACh in the brain. This line of research has produced some useful drugs, and, while they can slow the progress of the disease, they cannot completely arrest the decline in the brain's capacity to produce ACh, and so eventually their effects wear off.

Drugs designed to arrest or even prevent AD can be tested on a special strain of mice that have been engineered to accumulate amyloid plaques in their brains. The first drug specifically for the treatment of AD was tacrine, which was approved for use in the United States in 1993 but not released in the United Kingdom because of serious side effects. A better version was produced by the Japanese company Eisai, and called donepezil. This was launched in the United States in 1996 and in Britain in 1997, to be followed by rivastigmine in 1998. Tacrine and donepezil are treatments for mild and moderate AD, but rivastig, even appears to increase the level of ACh in those parts of the brain most affected by the disease.

Better still is galantamine, which became available in 2000, a drug that was discovered in the bulbs of snowdrops and daffodils. It improves mental activity and counteracts behavioural difficulties. Some people given galantamine even show signs of recovery, not merely just a slowing down of the condition. Galantamine not only blocks the breakdown of ACh, it also stimulates the brain's nicotine receptors, which in turn stimulate neurons to release more ACh. Nicotine is also thought to protect against AD because it can mimic the effects of ACh. Newer drugs are also on their way: especially promising are those that boost the nicotine receptors. As yet they are merely code numbers, such as GTS-21, SIB-1553A, and TAK-147, although some already have generic names such as nefiracetam and huberzine; the latter was first extracted from a Chinese herb, *Huperzia serrata*.

Another line of research has concentrated on counteracting metal toxicity, not of aluminium but of copper. It has been found that copper in the blood makes AD worse. This metal is essential for all living cells in the body, but clearly we must not have too much. The drug clioquinol can reduce the amount of free copper in the body by making it easy for it to be eliminated, and studies on mice with AD show that it can even remove plaques in their brains. Some metals appear to attract amyloid peptides and so begin the formation of plaques. Zinc appears to have this capacity, and clioquinol also binds to this metal and renders it inactive.

The most promising approach to treating AD would seem to be to prevent the formation of amyloid proteins, as this would then prevent the cascade effects of tangle formation that lead to inflammation, neuron destruction, and the outward symptoms of the disease. A team headed by Mark Pepys of the Royal Free & University College Medical School, London, found that a derivative of the simple molecule pyrrolidine-2-carboxylic acid has the ability to dissolve amyloid fibrils by reacting with a serum component that they rely on to keep them stable. The molecule reacts with this and removes it via the liver, and the unprotected fibrils then begin to break up. The team reported their findings in *Nature* in 2002.[45]

[45] Volume 417, page 254.

The active agent was discovered when Pepys tested the thousands of molecules in the collection of the pharmaceutical company Roche to see if they would bind to the serum component. Compound number Ro-15-3479 tested positive and, knowing this, Pepys then devised a similar compound that was even more effective: CPHPC.[46] It was non-toxic, well tolerated, and highly potent in low doses in removing the serum component from the bloodstream. As the level of this component drops, the decrease is registered and the serum component which is protecting the amyloid deposits then dissolves into the bloodstream, thereby exposing the deposits to attack by protein-recycling enzymes. Tests on mice were so successful that CPHPC was soon tested on human patients in the terminal stages of AD; it also proved to have a beneficial effect although it could not save them.

Finally, in *Nature Neuroscience* in May 2002, a team headed by Kelly Bales of Eli Lilly Research Laboratories published a paper reporting that a single injection of an antibody called m266 was able to improve memory in mice that had Alzheimer's. Whether this would happen to human sufferers of the disease is not yet known.

A urine test for AD is now available in the United States. AlzheimAlert detects neural thread protein, which is present only in those with the disease, and its level increases as the condition progresses and brain cells die. The pharmaceutical company Nymox produces the test, which involves taking a 50-ml sample of urine and submitting it for analysis to the company's labs in New Jersey. The cost is $295 and results are known within five working days of the sample being received. The company claims a success rate in detecting the disease of 90%.

There is now every reason to believe that AD will become a preventable, and even treatable disease, within the lifetime of readers of this book. This, too, may have a side effect in a way as yet unsuspected: older smokers who currently rely on nicotine to ward off AD will no longer have an excuse to continue with the habit.

---

[46] Its full chemical name is R–1–[6–[R–2–carboxy-pyrrolidin–1–yl]–6–oxohexanoyl] pyrrolidine–2–carboxylic acid.

CHAPTER SIX

Polymers in Disguise

IN THIS CHAPTER we shall be looking at four **polymers** (see Glossary) that play various roles in our lives, but we probably don't think of the products that contain them as coming from the chemical industry: disposable nappies, chewing gum, whispering asphalt, and CDs. The polymers on which they depend are polyacrylic acid (also known as polyacrylate), SBR (short for styrene butadiene rubber), SBS (short for styrene butadiene styrene), and polycarbonate.

Perhaps one day the names of these materials will be as familiar as the names of other polymers in our lives, such as nylon, polyester, PVC, and Teflon, and perhaps they could still be being used a hundred years from now. Whether the industry that makes them will still be around is debatable because chemiphobia continues to infect more and more people. That is an issue I shall deal with in the next chapter. For now, let me explain the contribution that four polymers make to our lives and why they are so successful.

Disposable unmentionables[47] and superabsorbent polymers (SAPs)

Although we prefer not to think about the emission of waste products from the human body, it is a problem for those people who have little or no control over their loss or disposal—namely, babies, menstruating women, the incontinent, and perhaps even space travellers. Thanks to a rather remarkable polymer, all these people can lead more comfortable lives.

---

[47] Nappies (diapers in the United States), incontinence pads, and feminine hygiene products.

A mother may derive a lot of pleasure from feeding her baby but she probably derives almost none from changing its nappies, and its father even less. During the course of a day, a baby will pass around 500 ml of urine plus between 10 and 50 grams of faeces. How this is dealt with is not something we tend to discuss in polite, or even impolite, society. Nor is it a topic that features strongly in the historical record. What writers down the ages have chosen to ignore chemists eventually confronted, and came up with a solution that most modern parents now use: the disposable nappy. Solving one problem, however, has led to another, as we shall see, although chemists are now working to solve that one as well.

Primitive people who lived in temperate or northern climes needed to wrap up their babies to keep them warm, and in doing so they presented themselves with the problem of how to deal with the somewhat random way in which their offspring urinated and passed bowel motions. All kinds of natural materials were pressed into service to absorb the unwanted mess. The Inuit probably had the hardest task in finding suitable materials to use but they solved the problem with moss, which they padded into sealskin diapers. In less remote regions, mothers would use grass or hay inside a soft skin such as that of rabbit, while in towns and cities cloth nappies were used, although these required frequent washing and had to be changed each time the baby soiled them if nappy rash was to be minimized.

With the Industrial Revolution came better materials, and especially cheap cotton towelling squares which made good absorbent nappies. Cleaning, washing, and drying these items became a major task in raising a baby. Things were only marginally improved in the 1940s with the introduction of plastic pants to cover the nappy to prevent leakage, and soft paper tissues which could be placed inside the nappy so that solid mess could be contained and flushed down the lavatory. These years also saw the appearance of the first disposable nappies, which consisted of a thick wad of paper tissues inside a plastic outer cover, but their capacity was such that they could take only one discharge of urine. They were not popular and not a commercial success.

The in 1961 the Procter & Gamble company of the United States introduced Pampers, which contained crêped cellulose fibres made from wood pulp and were much more absorbent. They were rather bulky, however, so much so that doctors complained that their size would force a baby's legs too far apart, and they warned that this could affect the development of its bones. In 1976 Kimberley Clark of the United States introduced hourglass-shaped nappies, Huggies, which countered this criticism (although today it is thought that bulky nappies may even be of benefit in that they encourage stronger growth of the hip joints).

Liquid retention is at the core of an absorbent pad. Materials that will absorb and retain water may do so by virtue of attracting the water to charged atoms (ions), or by being able to form a type of weak link to water, a link known as a **hydrogen bond** (see Glossary). Water is quite willing to form hydrogen bonds, and is attracted to the many oxygen atoms that are present in cellulose; this has several hydrogen-carrying oxygens and they, too, can hydrogen-bond to water. This is the reason why cellulose absorbs water easily, and why we use it in the form of paper for handkerchiefs, toilet rolls, and kitchen towels, and in the form of cotton and linen for table covers, hand towels, and bath towels. It explains why terry towelling makes excellent reusable nappies. Even sawdust, which is largely cellulose, was used as the absorbent material in old-fashioned incontinence pads for the same reason.

While water will cling to cellulose via hydrogen bonds these are not very strong, and can easily be severed and the water released again. Consequently, this absorbed water can easily be transferred to other things that the wet cellulosic material comes into contact with, such as skin. Sit a baby in a traditional wet and/or soiled nappy for too long and the result is invariably nappy rash in the form of red, raw, sore skin that is clearly painful. This can be soothed by applying oils or lotions, but the condition may be slow to heal because the baby must keep wearing nappies, and nappies are bound to get wet and dirty again.

Today nappy rash is rare, and the reason is mainly due to disposable diapers, which make nappy changing quick, easy, and hygienic. For most parents of a young baby the main drawback is the cost, and indeed the average baby uses around £2000 ($3000) in disposable nappies before it is toilet trained.

It was in the 1980s that a radically new type of disposable nappy appeared, and this was to revolutionize the nappy market. Superabsorbent polymer (SAP) replaced the thick wad of pulp fibre, making nappies thinner and neater, and with the capacity to absorb greater quantities of moisture. SAP had by then proved its worth by revolutionizing the female hygiene market, replacing sanitary towels with unobtrusive pads. Soon Velcro fastening tapes were added to nappies, thereby making the operation of changing them much easier. It was possible for a nappy to cope with several urine 'insults', and when the baby dirtied the nappy it was simply removed, rolled up, and thrown away. Its ultimate fate was either incineration or disposal to a landfill site, depending on the local refuse system. Before we look a little more closely at that worrying problem, let us examine the chemistry behind the diaper itself.

SAP is a rather remarkable polymer that can bind vast amounts of water, swelling as it does so. Indeed, the 10 grams of polyacrylic acid at its core can hold fifty times its weight in water, or thirty times its weight if put under the pressure equivalent to the weight of a baby sitting on it.

Polyacrylic acid was first recorded in a German science journal under the name of *Kunststoffe* (that is, 'Synthetic Materials') in 1938. This reported the research of a chemist, W. Kern, who had polymerized an aqueous solution of acrylic acid and obtained the polymer. A modified polymer can be produced by starting with a partly neutralized acrylic acid, so that there are acid groups and sodium acrylate groups present in the solution and the product is the more useful polyacrylate (see **polymers** in Glossary). As a result, some of the groups attached to the carbon chain are salt groups, which are ionic, and this greatly increases the ability of the polymer to absorb water, for reasons explained below.

In the commercial product, the polymer also includes a small amount of a cross-linking agent, which is necessary to make the polymer insoluble in water. (The more cross-links there are, the more insoluble the polymer becomes, but as it becomes more insoluble it becomes less able to swell, and so less able to absorb water.)

Patents for superabsorbent nappies and medical products based on SAP were first taken out in 1968, but it was not until the early 1980s that the first disposable SAP diapers were marketed in Japan, where they were an instant success. By 1984 they were on sale in the United States, and a couple of years later they appeared in Europe. They are now sold all around the world, to the extent that they consume a significant proportion of the 3 million tonnes of SAP produced every year. (The United States manufactures around 20 billion disposable nappies a year.) Other outlets for SAP are in adult incontinence items, an area that is rapidly growing, and feminine hygiene pads.

Many chemical companies manufacture SAP, and the largest manufacturer is BASF, which accounts for a quarter of global production; North America's share is around 40%, and Europe's is 30%. Most SAP has to be made close to where acrylic acid is manufactured, because this raw material is very reactive and tends to go off rather quickly, even when special inhibitors are added to preserve it. Consequently, there are scores of SAP manufacturing plants around the world, with the largest being able to manufacture more than 500,000 tonnes a year, although most have capacities of around a tenth of this.

Polyacrylate has other uses. Blending it into rubber produces a material that can be used as a sealant to prevent the entry of water into a structure. As it absorbs water it swells, and as it swells it forces the rubber polymers closer together, making a watertight membrane that prevents further penetration. The walls of the Channel Tunnel are protected in this way. Flexible water-blocking tape that can be wound round a leaking pipe is of similar composition.

Granules of SAP are used to counteract dampness in confined spaces such as wardrobes and cupboards, where high humidity can lead to musty smells, and in kitchen waste-bins, where microbes can breed. When the SAP has become saturated, it can be heated in an oven to release the absorbed water and used again and again. A very different use is in horticulture. A handful of SAP granules is mixed with potting compost in which seeds or seedlings are planted, and this ensures that more water is retained so that they thrive better. Several commercial products are available for situations where water is likely to be of irregular or limited supply, such as in window boxes and hanging baskets. These are relatively minor uses of SAP compared with the disposable nappy market.

A disposable nappy has four components:

1. The body-side liner, which is made from polypropylene fabric because this feels soft against the skin. This layer also has two flaps running along its length, and these act as a barrier to prevent urine flowing outwards.

2. The surge layer, which is designed to disperse the urine from its point of impact. This layer is made of cellulose or a polymer mix.

3. The all-important absorbent layer, which consists of a mixture of the SAP dispersed in cellulose fluff, all contained in a layer of tissue.

4. The outer covering, again made of waterproof polypropylene with a cloth-like feel. This layer also has built into it the stretchy ears which fasten the nappy, and elastic at the outsides to grip the legs, thereby preventing excess urine from leaking on to clothes or bedding. This outer layer may also be made permeable with millions of tiny holes, allowing air to circulate to reduce the steamy atmosphere inside the nappy, thereby helping to keep the baby's skin drier.

SAP works on the principle of **osmosis** (see Glossary), whereby water will tend to move through a membrane from a less concentrated solution to a more concentrated solution until the concentrations at either side are the same, when osmosis ceases. Water moves into the SAP in an unsuccessful attempt to dilute the concentration of sodium and acrylate ions, and if pure water is in contact with the SAP, then it will continue to be absorbed to such an extent that as much as 3000 times its weight of water may be incorporated, yielding a colourless gel. In a nappy this does not happen, because urine is not pure water, but more like a 1% salt solution. As the water transfers to the SAP, the urine itself becomes more concentrated until equilibrium is reached; that is why a nappy will at best absorb only fifty times its own weight of urine, but this is more than enough.

As might have been expected, disposable nappies were not acceptable to some environmental groups who foresaw the problem of their disposal. They campaigned against them in other ways, however. 'Disposable nappies linked to male infertility' was a typical headline in newspapers across Britain and the United States in September 2000, and the stories were based on a paper in the *Archives of Disease in Childhood* reporting the research of a team led by Professor Wolfgang Sippell of the Department of Paediatrics at Kiel University, in Germany. This had involved forty-eight baby boys whose mothers used disposable nappies, and the study had found that, on average, the temperature of their scrotums was a degree higher than those of baby boys wearing the washable terry-cotton type of nappy.

Because sperm production is inversely proportional to the temperature of the testicles, that is, the higher the temperature the less developed and mobile the sperm, the finding allowed researchers to speculate that boys who wore disposable naappies might well suffer problems later in life. In the media this research was also linked to the much-publicized decline in the fertility of Western males in the second half of the twentieth century. Clearly, if the theory was correct, the sooner disposable nappies were banned the better.

These alarming news stories preyed on parents' concerns about their male offspring's ability to father children, and that's what made the report newsworthy. Because it appeared in a respectable journal, it had to have been scrutinized by other learned paediatricians before it was published and so could not be dismissed merely as scaremongering. Undoubtedly, the wearing of disposable nappies was causing a one-degree rise in the temperature around the genital area of boy babies. That would seem logical, because at the time the study was carried out, the back sheet of a disposable nappy was not only waterproof but effectively sealed in the warm, moist air as well. Boy babies wearing cotton nappies would not experience this effect, assuming their mothers did not put plastic pants over the nappies. With the freer circulation of air there would be a mechanism for cooler air to move within the nappy region.

The second attack on disposable nappies could be taken more seriously, and it came from those who were worried about their non-biodegradability, which meant that used nappies going to landfill would take centuries to decompose. Reusable cotton nappies, therefore, should be favoured over disposable nappies because they reduce the amount of such waste. Indeed, these are welcomed and even encouraged in some places, despite the fact that eco-analysis has shown that they offer little benefit in terms of overall energy consumption and use of resources.

More serious is the criticism that disposable nappies are dependent on the petrochemical industry, and as such they are ultimately derived from fossil fuels, which might make them unsustainable in the long term.

These challenges are ones that we must address. Can we produce disposable nappies that are both biodegradable and come from renewable resources? The answer is yes, although the earliest attempts to do so had limited success.

A Swedish mother of two young boys, 42-year-old lawyer Marlene Sandberg, saw that existing disposable nappies could be made much more environmentally acceptable, and she created one that is sold under the brand Nature Boy & Girl; these are made from 70% biodegradable and renewable resources, with cotton providing the bulk of the absorbency layer, and a biodegradable polymer derived from maize providing the plastic outer layers. Of course the nappies still had some SAP, but much less than conventional disposable nappies. While such a development was to be welcomed, these nappies still fell foul of the Women's Environmental Network, whose spokesperson was quoted as saying 'a disposable nappy, whatever it is made of, is contributing to a large waste problem'. Moreover, the established makers of nappies discount Sandberg's claims, maintaining that, if her nappies end up in a landfill they won't disappear. They say that her nappies are more to do with clever marketing than clever science—and to a certain extent they are right.

There is no reason why, one day, all the components of a disposable nappy should not come from renewable resources. They may then be disposed of by biodegradation in the tanks of methane generators, or composted to make fertilizers, or even burnt in incinerators to generate heat, although their high water content would suggest this is a less likely method of disposal. Another solution to the disposal of disposable nappies is being tested in the Californian city of Santa Clarita, where there is roadside collection of used diapers. These are then sterilized and separated into paper and plastics in a unit that has a throughput of a tonne an hour. The fibre pulp will be used to make paper, and the plastic to make outdoor furniture and boarding.

A Mexican company, Absormex, claims that its Natural Baby Supreme diaper is entirely biodegradable, and says it has achieved this by adding an agent to the plastic layers which encourages their decomposition by sunlight, heat, or mechanical action. In fact, they are so sensitive to the ultraviolet rays of the sun that they have to be sold in packaging that protects them from UV.

Meanwhile, chemists continue to improve existing disposable nappies. One problem with SAP was that it is not always successful at absorbing all the urine, because its particles can form a non-permeable layer on the outside of the particles, preventing them from working properly. Creating SAP with a better surface area that was less prone to gelling soon solved this problem. Another version of SAP may soon come onto the market: this is based on polymethacrylate gel, which is not only a better absorbent than ordinary polyacrylate gels, but also acts as a deodorizer as well, which is a key factor in the care of elderly people.

Other companies are also working to find ways to neutralize the odorous thiols that are present in urine, and which give it its characteristic smell.

Chewing gum: polymeric hydrocarbons: 1

Chewing gum was once made entirely from natural sources; today it relies very much on synthetic materials, and as a result, is not only a better product but also much healthier.

Communication between people who come face to face involves not only the spoken word, but also gesture, body language—and smell. Nothing interferes more with the desire to continue a conversation than halitosis. What can we do to prevent this, or to remedy the situation? One answer is to rinse your mouth with Listerine or some such mouthwash, but this may not be practicable during the course of a working day. The other answer is to help the mouth refresh itself by chewing some gum.

Some smells come from the food we eat, such as raw onions and garlic: the sulfur molecules they contain are expelled on the breath. This situation may last a few hours but we are generally aware of the odour ourselves. If, however, such molecules are being generated by bacteria in the mouth breeding on food debris, or living in tooth cavities, or on diseased gums, we may be unaware of their presence on our breath, and without our knowing it they may be seriously affecting our professional and social life.

A part of human grooming is oral hygiene and, while you may clean and even floss your teeth after breakfast, this is unlikely to provide the protection you need throughout the day. The simplest way to freshen your breath is to chew gum. This will not only add its own minty flavour to your breath, it will also remove food particles and bacteria from the teeth, and by stimulating the flow of saliva it will wash the mouth.

Modern chewing gums may be the sophisticated product of food chemistry but they are still far from ideal—one major problem is the disfigurement of the urban environment by discarded gum. In 2000, a survey of the pavements and road in London's main shopping street,

The chemistry of saliva

The average healthy person secretes half a litre (500 ml) of saliva a day. The rate at which the six glands in the mouth produce it varies from as low as 0.3 ml per minute when the body is at rest, to as high as 5 ml per minute when chewing first starts, although this falls to around 1.5 ml per minute after 20 minutes.

Saliva is 99.5% water, but the dissolved chemicals in the other 0.5% play a crucial role in oral and dental health, as well as their obvious one of providing digestive enzymes. The chemicals in saliva consist of organic molecules, inorganic ions, and macromolecules.

The organic molecules are, in decreasing order of relative amounts: fatty acids, urea, free amino acids, uric acid, lactate, glucose, and their relative abundances roughly reflect that of the body's plasma, albeit at much lower levels.

The inorganic ions are, again in decreasing order of importance, chloride, potassium, sodium, phosphate, bicarbonate, calcium, and magnesium. The calcium and phosphate are essential to keeping the teeth in good condition, because tooth enamel is calcium phosphate, and in this way, the saliva helps repair the teeth. The pH of saliva is effectively neutral, being 7, and it is designed to promote remineralization of the tooth enamel. It can do this provided the pH does not fall below 5.5 , at which point slight demineralization occurs.

The macromolecules are proteins, glycoproteins, antibodies, and lipids, plus the enzymes *amylase, peroxidase,* and *lysozyme* (which is an antibacterial agent). The number of proteins detected in saliva has increased considerably in recent years and we now know of fifty, although some of their functions are still unclear.

Lack of saliva is known as xerostomia, and this is a symptom of many conditions, including being a side effect of some common drugs. Chewing gum is an ideal remedy.

Oxford Street, showed that there were more than a quarter of a million blobs stuck to it. The government of Singapore outlawed chewing gum in 1992 because it was interfering with the rubber edges of the sliding doors on underground trains. (They partly lifted the ban in 2002 by allowing sugarless chewing gum to be sold in pharmacies, provided that a doctor or a dentist had prescribed it.) Chewing gum sticks so well to asphalt and rubber because it is a case of sticking like to like: asphalt, rubber, and chewing gum are made of polymeric hydrocarbons.

In theory, it should be possible to make a chewing gum that does not deface public places. An obvious answer to the problem would be to make it edible, so that when we have finished chewing we could swallow it discreetly, rather than have to spit it out and throw it away. Alternatively, it could perhaps be made biodegradable, although that would probably be difficult, because it would have to survive unchanged in the mouth while exposed to saliva and digestive enzymes, and yet be quickly susceptible to other enzymes in the environment. No doubt public cleansing departments will have to continue to wrestle with the problem because chewing gum is particularly persistent.

People have chewed gum for thousands of years, and clearly the gums they chewed must have come from trees which ooze such gums when their bark is damaged. In 1993, Bengt Nordqvist, of the Swedish National Board of Antiquities, came across the oldest piece of discarded chewing gum ever discovered when he was investigating a prehistoric dwelling, a 9000-year-old hut on the island of Orust. That it was chewing gum was proved by the fact that it still had teeth marks imprinted in it. The gum was the resin from a birch tree, in which case it was not unlike some modern 'sugar-free' gums because it would, like them, be sweetened with xylitol, the more common name of which is birch sugar. It was more than likely that a Stone Age teenager had discarded the gum, because the marks showed his or her teeth were in a healthy state.

The ancient Greeks certainly chewed gum. The physician Dioscorides, who lived in the first century AD, recommended chewing mastic for its curative powers. This is the aromatic resin which exudes from cuts in the bark of the mastic tree (*Pistacia lentiscus*), which is indigenous to the Mediterranean region, growing well in coastal districts and particularly on the Greek islands. In Roman times, the production of mastic resin was the exclusive export of the Greek island of Khios, although most of it went into making varnish. In the summer months, the mastic tree was cut with lots of vertical incisions, from which sap oozed and slowly hardened to form large oval drops the size of peas. These were harvested every 15 days between June and September.

The modern habit of chewing gum can be traced back to the Native American Indians of New England, who chewed the gum from the spruce tree, and this, too, was obtained by cutting gashes into the bark. By the early 1800s, the American settlers had taken up the chewing-gum habit, and in 1848 John Curtis of Bangor, Maine, started selling it commercially as 'State of Maine Pure Spruce Gum'. Two years later he moved to Portland and changed its name to American Flag. He also made a chewing gum from paraffin wax and sugar, which he sold under names such as Sugar Cream gum and White Mountain gum.

The biggest breakthrough in chewing gum came with the introduction of chicle, a latex gum from the sapodilla tree (*Achras zapota*), which grows in the forests of Yucatán, Guatemala, and other regions of Central America. The Americans were only rediscovering the chewing gum once enjoyed by the people of the Mayan civilization, which flourished up to about AD 800.

Tradition has it that General Antonio López de Santa Anna (1797–1876), former President of Mexico, took some chicle to New York and gave it to a Staten Island photographer named Thomas Adams, suggesting that it might be used as a substitute for rubber. It proved impracticable for this purpose, but Adams decided it might be used as a chewing gum and, together with his 12-year-old son Horatio, he turned it into strips and marketed it as Adams New York Chewing Gum, selling at 1 cent per stick. The year was 1871. Chicle was an ideal chewing-gum base; it gave a smooth, springy chew and it also retained flavours well. It was to be the main ingredient in chewing gum for 50 years, until it was displaced in the 1940s by synthetic polymers. The disadvantage of chicle was that it had to be harvested from individual trees growing in the jungle. The tree was unsuited to being grown in plantations, despite attempts to do so.

At one time the United States imported 7000 tonnes of chicle a year from Central America, but today it is less than 200 tonnes. Perhaps this is just as well: the harvesting can be done only from trees that are at least 20 years old, and then they will yield only about a kilogram of gum per tapping, and this can take place only once in 3 or 4 years. The sap flows best at night and in the early morning. The milk is gathered and then heated, which causes it to polymerize into a sticky mass, when it is poured into wooden moulds to cool and harden. Harvesting chicle is fraught with danger, not only from poisonous snakes that live in the forest but also from the chicle fly; this will lay its eggs in a tree-tapper's nose and ears, producing grubs that eat their way into the flesh and cause facial deformities. Moreover, the gum sells for only about $2 per kilogram, and it accounted just a for few per cent of the value of the chewing gum that was once made from it.

Modern chewing gum is made from synthetic elastic polymers (elastomers), and these are still a minor component, at least in terms of what you pay. The price of a packet of chewing gum is made up as follows: raw materials 20%; manufacturing 25%, distribution 10%, sales and promotion 30%, and profit 15%.

In addition to chicle, the natural gums that were once used came mainly from tropical plants with names such as chiquibul, jelutong, perillo, sorv, and tunu. (The latex from rubber plantations was no use because it has the wrong texture.) These gums are derived from isoprene, a simple hydrocarbon molecule produced by the trees. Isoprene is a volatile liquid, boiling point, 34 °C, with the chemical formula C_5H_8. The molecule consists of a chain of four carbon atoms, with a double bond at each end and a methyl (CH_3) group attached to one of the middle carbons. Isoprene, like all such double-bonded materials, will polymerize under the influence of catalysts, and in the natural state, it is the oxygen molecules of the air that trigger this process. One can speed this up by gathering the sap, acidifying it, and boiling it, whence it coagulates into a soft, rubbery mass. The polymerization produces long hydrocarbon chains that are soft, springy, and chewable.

Natural gums had a major disadvantage in that they could have quite a strong flavour. Ideally, the gum base should have no taste of its own, because that then allows a wide range of more attractive flavours to be added, so chewing-gum manufacturers prefer to use material made from pure isoprene. This chemical is produced on a large scale by the petrochemical industry for the manufacture of various synthetic rubbers.

Once the Rubicon, from natural to artificial elastomers, had been crossed, there was clearly no reason to limit the choice of elastomers only to polymers of isoprene. Today, therefore, we find that other materials are used, including polyisobutylene, polyvinyl acetate, polyvinyl laurate, and especially co-polymers of butadiene with styrene. Exxon Mobil Chemicals, part of the giant oil company, produces high-purity polymers

for chewing gums. Vistanex LM is a low-molecular-weight polyisobutylene, whereas Vistanex MM is a high-molecular-weight polymer and is
supplied with antioxidant added. The company also supplies Butyl 007,
which is a co-polymer of isobutylene with small amounts of isoprene.

Synthetic gums were first introduced in the United States because of
the increasing scarcity of those of plant origin, and they are the only ones
now used in chewing gums. Those approved for use in chewing gum
include styrene-butadiene rubber (SBR), isobutylene–isoprene copoly-
mer, paraffin wax, petroleum wax, polyethylene, polyisobutylene, polyvinyl alcohol, and synthetic terpene resin—see **hydrocarbon polymers** in
the Glossary for more information about these.

The world's best-known company making chewing gum is still that
which Adams founded and which, in 2002, became part of the Cadbury
Schweppes group of companies. Another famous chewing-gum manufacturer is Wrigley, which was founded by William Wrigley in Chicago in
1892. He started out in business selling first soap and then baking powder, and to boost sales of the latter he gave away packets of chewing gum
with every purchase. This had the desired effect, but Wrigley soon realized that there was more profit to be made in selling chewing gum. His
biggest seller, Wrigley's Spearmint Gum, was launched in 1906; within
four years it was the best-selling brand in the United States, which is still
the world's largest market for gum. On average, Americans each chew
300 sticks of gum a year, at a total cost of more than $2 billion. The area
of farmland used to grow the mint to flavour such gums covers around
150 square kilometres, in states such as Idaho, Oregon, and Wisconsin.

In Europe, more than eighty ingredients are permitted in chewing
gum, and about fifty in the United States. The average chewing gum contains at least fifteen, and often twenty, ingredients; some even have thirty. The essential ingredients in all chewing gums are the gum base,
sweeteners, flavours, emulsifiers, humectants, and preservatives. While
these are listed in order of importance for ensuring a good product, they
are not the order in which they make up its components. In a traditional
stick of Wrigley's Spearmint Gum the main ingredient is sugar, which
accounts for about half its weight. Then comes the gum base, which
makes up about a third, while the remaining ingredients, including a
little water, generally account for 1 or 2% each.

Rubbers become softer when they absorb oil, and the same thing happens to gum bases when waxes are added; these make the gum more workable because they act as lubrication between the strands of polymer. Waxes also consist of chains of CH_2 groups linked together, and natural waxes were mentioned on page 4. Those from the petrochemical industry are more standardized. The waxes with shorter chains (25 to 30 carbons) are called crystalline wax, and the ones with longer chains (35 to 50 carbons) are called microcrystalline wax.

Natural waxes used as plasticizers include carnauba wax and beeswax, and these are blended to get the right consistency and to control the rate of flavour release. Petroleum waxes are better in that they give a longer shelf-life to the product. The crystalline waxes are given EU food-additive numbers, such as E907 for crystalline wax and E905 for microcrystalline wax, although the latter tends not to be used, following the accusation by one food pressure group that it had caused cancer in laboratory animals. (Proof that it was carcinogenic was never forthcoming, but the adverse publicity meant that this was no longer acceptable as a food additive.)

Other important ingredients in chewing gum are emulsifiers such as lecithin and glyceryl-monostearate, which are needed to soften it and to enable the various components to blend together to form an homogeneous mixture. Humectants are included to prevent the gum from drying out and going hard. The best humectant is **glycerol** (see Glossary). Emulsifiers and humectants together constitute less than 1% of the product.

Various sweeteners are added to chewing gums, the traditional one being icing sugar, which is sugar ground to a fineness of 200 microns or less. Other sweeteners include corn syrup, sorbitol, mannitol, xylitol, aspartame, and acesulfam. Corn syrup is the best because it improves the texture of the gum, but it is counter-productive in terms of dental hygiene. Like sugar, it can feed the *Streptococcus mutans* bacteria in the mouth, and these produce acids which can eat through tooth enamel and thereby cause cavities. Xylitol, aspartame, and acesulfam sweeteners are much safer, and, even though xylitol is a carbohydrate like sugar, unlike sugar it is not fermented to acid, and so provides a safe form of sweetness. Aspartame and acesulfam are intensely sweet and so only a tiny amount is needed, but in any case, these artificial sweeteners cannot cause tooth decay.

Flavours account for about 1% of a chewing gum, and the most popular are spearmint, in the form of the chemical carvone, and pepper-mint, which is mainly menthol. Fruit esters are sometimes used for fruit-flavoured gum. The average chew lasts about 20 minutes, and in modern chewing gums, the flavour will last throughout. In Japan, there used to be a novelty gum called Mystery Gum, which changed flavour as it was chewed so that at different times it would taste of peach, grape, pineapple, and strawberry.

A typical preservative used in chewing gum is BHT (short for butylated hydroxytoluene, EU food-additive code E321) which is a powerful anti-oxidant and so protects the chewing gum from attack by oxygen. Only about 100 p. p. m. (0.01%) of this is needed to do the job. In some coun-tries, this preservative is not permitted and other antioxidants, such as alpha-tocopherol, have to be used. Antioxidants also had to be added to the natural gums when they were used to make chewing gum because the double bonds in their polymer chains were just as susceptible to chemical attack by oxygen from the air, which leads to cross-linking and thereby hardens the gum. (These antioxidants replaced the naturally occurring antioxidants which were removed during the washing and purification of natural rubbers.)

Chewing gum is made in large vats, in which a tonne of gum is melted and the other ingredients are added. The whole is then stirred until it has the consistency of bread dough, and for traditional chewing gum it is then rolled into thin strips, dusted, cut into pieces, and wrapped. Chew-ing gum can also be passed through rollers that cut and press it into small blocks which are then given a sugar or xylitol coating, the latter if it is to be sold as sugar-free.

Chewing gum can contain active substances that might have health benefits, such as fluoride for strengthening tooth enamel, vitamins, and one kind even contains p-chlorbenzyl-4-methylbenzylpiperazine, which prevents travel sickness. Such active ingredients generally make up to 5% of the product. It has been suggested that other medicaments could be delivered this way, but this method of dosing has not been approved. The idea of medicated chewing gum as a slow-release prescription is theoretically possible by blending it with polyvinyl acetate (PVA) and incorporating this into the gum as a plasticizer. PVA is already used as a binder in pharmaceuticals, and it would ensure that the drug was re-leased only slowly.

Oral sex?

A patent for Viagra chewing gum was filed by the Wrigley company in November 2000. Each stick of the gum will contain between 5 and 100 mg of sildenafil citrate, and the user would chew it about half an hour before he wanted to engage in sex. Not that Viagra chewing gum is likely to be available until after 2013, when the Viagra patent runs out and cheap generic versions of the drug can be obtained. The benefit of Viagra chewing gum would come from the slow release of the active ingredient, thereby overcoming the stomach irritation that some men experience when taking the pill.

Chewing gums can be made to do other things, such as not stick to dentures, make teeth whiter, and unblock congested nasal passages (by including menthol or eucalyptus). Freedent is chewing gum for those who have dentures or crowns, to which ordinary chewing gum tends to stick. Its gum base is less tacky.

Bubble gum has a gum base with added elasticity, making it possible to blow bubbles with it. It was invented by an American, Frank Fleer, in 1906 and he called it Blibber Blubber, but it was not popular because it easily burst before a satisfactory bubble could be blown, and, what was worse, it stuck to everything it touched. His firm produced a superior variant, Dubble Bubble, which went on sale in December 1928 and is still sold today. Bubble gum's gum base is either SBR or just butyl rubber.

In 1996, a US company, Church & Dwight, test marketed a therapeutic chewing gum that reduces plaque. It was based on 'baking soda', which neutralizes mouth acids (baking soda is sodium hydrogen carbonate, $NaHCO_3$, also known as bicarbonate of soda, or simply bicarb). According to the company, the gum should be chewed after brushing the teeth and it will reduce plaque by as much as 25% within a month of regular use. To make it palatable, the taste of the sodium hydrogen carbonate was disguised with a combination of sweeteners, including xylitol, and flavours such as peppermint, spearmint, wintergreen, and cinnamon.

Whispering roads: polymeric hydrocarbons: 2

There is much less traffic noise than there used to be, and this is not just because of quieter engines and sound-absorbing insulation; tyre noise, caused by the rubber of the tyres striking the road surface, is a major component of traffic noise. This, too, can be greatly reduced by changing the nature of the asphalt used in building the road, and at the same time this reduces the spray in wet weather. The secret is to modify the bitumen in the asphalt by adding polymers.

Dirty, tacky, sticky, and messy—that's tar. Indeed, we talk of being 'tarred with the same brush', of being 'tarred and feathered', of being 'defiled by pitch', of the night being 'pitch black'. All are phrases that have been part of the English language for a long time, and all refer to a material that may have been useful but which had its downside if you, or your possessions, came into contact with it. Not only was it very dirty, it stuck like glue. Today, we talk not of pitch, or tar, but of bitumen. Pitch is the residue obtained by distilling natural wood oils such as turpentine; tar is the residue from the dry distillation of coal; and bitumen is the residue left behind when crude oil is distilled.

Pitch has a long history and was used for caulking the seams of boats. Tar was used mainly in road building, but this fell out of favour when it was realized that it could cause cancer. Today, the preferred material is bitumen, which is used for roads and also for roofing, and it is not carcinogenic. We may still not wish to have direct contact with it—and in that respect we've not changed—but the material we are talking about has undergone a revolution in the past twenty years. For example, it is no longer a case of its always being black; you can now have it any colour you desire.

Bitumen used to be the 10% of tar-like residue that was left over when all the useful hydrocarbons had been distilled from crude oil but, thanks to improved methods of refining it now accounts for less than 1% of most oils, though some crude oils yield half their weight as bitumen. Of the 1500 sources of crude oil around the world, only a few are suitable for bitumen production, yet demand continues to grow because it is essential for roads, runways, and roofs. Bitumen production is now undertaken at specially constructed plants in modern refineries, and total output world-wide is more than 75 million tonnes a year. What the chemists have done is find ways to improve it, and even make it environmentally friendly.

The chemical composition of bitumen is approximately C_7H_{10}, with small amounts of oxygen and nitrogen, but with as much as 6% sulfur. Not that this tells us very much, because it is a veritable witch's brew of chemical compounds. We do know, however, that most of the molecules in bitumen are large and non-volatile, otherwise they would have boiled off when the crude oil was fractionated.

There are two main types of ingredients: the asphaltenes, which are insoluble in the solvent heptane, and the maltenes, which are soluble. In effect, bitumen is a colloidal dispersion of asphaltenes in maltenes. The asphaltenes are the minor component and are large molecules. The maltenes are the major component and consist of aromatics, saturates, and polar molecules, in that order of importance. The aromatics are mainly derivatives of naphthalene, the saturates are viscous oils, and the polar molecules provide the adhesion, the proverbial stickiness for which bitumen is infamous.

Bitumen was no good as a source of raw materials for industry, yet it was quite satisfactory for conventional road surfaces. Demands for better roads, however, required a better type of bitumen, and this meant adding things to it. Bitumen itself is a thermoplastic material, which means that its texture depends on the temperature; the higher this is, the more flexible the bitumen is, and the more flexible it is, the more it deforms under the pressure of heavy traffic. Reclaimed rubber from tyres was once added to it, to counteract this tendency by increasing the strength of the bitumen and making it more resistant to being deformed. Adding sulfur and organomanganese compounds also made it more workable and even stronger, but the best way of modifying it is to add thermoplastic polymers, particularly SBS (styrene–butadiene–styrene).

Bitumen that has been modified by SBS has greatly improved performance, which makes it able to take more strain and not to deform when it becomes overheated. The benefits are measured in terms of the pressure it can withstand before it develops cracks, and modified bitumen is 100 times stronger than ordinary bitumen. SBS is produced by combining two polymers, with very different properties, and which are normally incompatible. There is a highly elastic rubber as the middle section of the polymer and this comes from the butadiene, with polystyrene at both ends of the chain, which is much harder and provides the strength. When SBS is added to hot bitumen, it swells up enormously, and then, as it cools, the polystyrene end-blocks link to form a three-dimensional network. SBS is primarily used in road and roofing bitumen while a similar material, styrene–isoprene–styrene (SIS), is used for hot melt adhesives and bitumen sealants.

Back to the Bible

Long-chain polymeric hydrocarbons have a long history. In the ancient world, pitch was used to make the hulls of ships waterproof, to seal water tanks and channels, as a mortar for bricks, and as an adhesive in the making of tools, weapons, mosaics, and inlay work. According to the Bible, Noah was told to build the Ark of gopher wood, and coat it with pitch both on the inside and the out (Genesis 6: 14). The Tower of Babel was built of bricks with bitumen as mortar (Genesis 11: 3). And the small boat made of bulrushes in which Moses was cast adrift on the Nile was also sealed with pitch (Exodus 2: 3). While these tales were mainly allegorical, their authors obviously knew of the ways in which bitumen was used in the ancient kingdoms. Natural bitumen seeped out of the ground in several places, and lumps of it were even to be found floating on the Dead Sea, then known as Lake Asphaltities, from which the term 'asphalt' is derived. Today, the word asphalt refers to a mixture of bitumen and other components.

The oldest known use of natural bitumen was discovered at Mohenjo Daro in the Indus valley, where archaeologists found a 5000-year-old water tank constructed from stone blocks that were bonded together with bitumen. (Bitumen is still used for this purpose in modern reservoirs.) The people of Mesopotamia, and the Chaldeans, Akkadians, and Sumerians mined bitumen from shallow deposits and exported it far and wide. We may imagine the use of bitumen in road building to be a relatively recent development, but bitumen was used for this purpose in the sixth century BC, in the days of King Nebuchadnezzar of Babylon, although back then it was used as a grout for paving stones.

It was once thought that Egyptian mummies got their name from *mumiyah*, the Arabic word for natural bitumen; the black appearance of the mummies was thought to be due to their having been coated with pitch. In 2001, chemists Richard Evershed and Stephen Buckley of the University of Bristol carried out a chemical analysis of thirteen mummies from museums in the United Kingdom. While they detected all kinds of resins and preservatives used by the ancient embalmers, there was not a trace of bitumen.[48] Egyptian mummies of the Roman era were sometimes covered with a layer of bitumen to make them waterproof, but this was never the practice in the days of the Pharaohs. Mummies then were coated with beeswax, and this might well be responsible for the name 'mummy' because the Egyptian Coptic word for wax is *mum*.

Natural bitumen is not as rare as we might imagine. The English courtier and navigator, Sir Walter Raleigh (1552–1618), found the famous Pitch Lake in Trinidad in the 1590s. This covers almost 50 hectares and contains more than 6 million tonnes of bitumen, which is being replenished from below almost as fast as it is removed and exported. An even larger pitch lake is that at Guanoco Lake, Venezuela, which is almost 500 hectares in area; bitumen was extracted from it from 1890 to 1935.

The La Brea tar pits, in Hancock Park, Los Angeles, California, were

---

[48] Their research was published in *Nature* in 2001, volume 413, page 837.

discovered in 1769 during an expedition led by Gaspar de Portolá, and were to become much more important for the things extracted from them. The bitumen contains the fossilized skulls and bones of prehistoric animals which had become trapped in the sticky ooze. More than a million prehistoric specimens have been found in its tar, including mammoths, mastodons, sabre-toothed cats, giant sloths, and even a camel.

The earliest-surviving photograph was recorded with the help of bitumen. Joseph Nicephore Niépce (1765–1833), pronounced 'Nee-yeps', took it in the summer of 1826 from an upstairs window of his home in Chalone-sur-Saône, Burgundy, France. He had been experimenting with light-sensitive substances since 1816, the year that he made his first crude images, but achieved better quality in the 1820s. The photograph shows a farm building and a pear tree, and it was captured on a copper plate covered with bitumen. On exposure to sunlight for many hours, and possibly over the course of several days, the jet-black bitumen not only faded to a light grey but also hardened. Niépce washed the plate with a mixture of oil of lavender and white spirit, removed the unaffected bitumen, leaving behind a permanent image imprinted in hardened bitumen of varying shades of grey. Niépce had used a material called Bitumen of Judea, which was a naturally occurring substance used in engraving processes. Although named after the area in the Middle East from which it originally came, it was also to be found in the south-east of France where Niépce lived.

Niépce called his image a heliograph, and brought some of his pictures with him to London when he came on a visit in 1827. He went to the famous gardens at Kew where he met the botanist Francis Bauer, who was a fellow of the Royal Society, and to whom he showed samples of his work. Bauer recognized at once the importance of what Niépce had done and persuaded him to submit a paper on his discovery to the Society that December. Although he wrote the paper, it was not published because he refused to reveal the methods that he had used. Niépce returned to France, but his method of photography was too slow ever to be popular. Fortunately his photo of the farm building and pear tree was not destroyed. It was rediscovered and authenticated in 1952, and is now on display in the Harry Ransom Humanities Research Center at the University of Texas at Austin.

Bitumen and modern roads

Supplies of natural bitumen were not enough to meet the demands of the industrial nations of the nineteenth century, and other sources became more important. These included the tar produced as a by-product of the gas-works that supplied towns in Europe and North America, and which used coal. The so-called gas tar was ideal for road building.

Today, all bitumen comes from oil. When crude oil is heated to 300 to 350 °C, all kinds of hydrocarbons distil off. Some of these are gases (such as propane and butane), but most are volatile liquids used for fuels (for example, petrol for cars, kerosene for aircraft, diesel for trucks). The distillation temperature is eventually increased to 400 °C and the pressure reduced to about half atmospheric, to remove more of the hydrocarbons. What is left behind is a residue that is ready to be used as bitumen. Other grades of bitumen are made by blowing air through it while it is hot. This 'blown' bitumen becomes partly oxidized, which makes it harder and more viscous because the oxygen atoms react to link the hydrocarbon chains together.

Much has been done to make roads more economical, safer, and even environmentally friendly, because the materials from which they are made can be recycled into new road surfaces. Coping with extremes of weather is equally important, as roads are built on high mountains and inside the Arctic Circle, where their surfaces may be frozen for much of the year. These need a softer type of bitumen than that used for roads which are exposed to the intense heat of the tropics.

Roads can now be made to take heavier loads, thereby reducing the transport cost of moving raw materials and finished goods; they can be made using less material; they can be made to last longer before needing repair or renewal;[49] and they can be environmentally friendlier by producing less noise pollution and requiring less lighting. Roads can be made safer by reducing the hazards associated with wet roads, such as spray and skidding, and they can be aesthetically more pleasing by being colours other than black. Coloured bitumen is possible when the base material is almost colourless, as it can be when produced from some

---

[49] Heavy vehicles damage roads the most; the passage of one 40-tonne, five-axle, heavy goods vehicle has the same effect as the passage of 500,000 cars.

types of oil. Indeed, there are red, yellow, blue, and green asphalts being used for promenades, cycle lanes, and bus lanes. In tunnels, it is possible to lay 'white' asphalt, and this has the added bonus of reducing the need for bright lighting.

Roads account for 80% of the bitumen that is produced, and, while bitumen accounts for only a few per cent of the asphalt that comprises a road surface, it is the essential component, the adhesive that holds the other components together. These consist of sand, aggregate, lime, and the rubber of old tyres, but whatever is used, a modern road should be built to last fifty years before needing to be reconstructed.

Bitumen is ideal, because when it is hot it is fluid and easily mixed, and when it cools it forms a hard, tough, durable, and flexible surface. SBS had been developed in the 1960s, and this elastomer seemed ideal for industries such as footwear manufacturers that needed strong adhesives. The firm that developed the material was Shell Chemicals and it called its product Kraton. When this was added to bitumen it enhanced its properties, especially its flexibility and elasticity, which greatly improved its use with roof coverings. It also raised the softening point of bitumen from its usual 50 °C to 90 °C, and this made it good for road surfaces; asphalt made with Kraton would not deform under the weight of traffic, even in very hot countries.

Between 3 and 7% of SBS is added to the bitumen, depending on its application, but the cost is reclaimed in a greatly extended road-surface life (up to five times longer than ordinary asphalt) and fewer repairs. It is ideal for bridges and airport runways, where stresses are especially high. Suspension bridges generally carry roads that are not only heavily used by traffic but also tend to be constantly in motion, bending and flexing according to the volume of traffic and the weather. Moreover, the road surface has to be as light as possible, and this means the asphalt has not only to be strong and crack resistant but also much thinner than on a normal road. The answer is to incorporate 7% of SBS.

Asphalt roads can be laid hot (100 to 140 °C is the usual range of temperatures at which asphalt is rolled) or cold (when the bitumen is emulsified and mixed with aggregates). A chemical reaction between the bitumen droplets and the aggregate surface then binds the two together. When wet road conditions are the norm, the bitumen needs to have about 1% of a surfactant (cationic) to prevent water damage.

One of the first road surfaces to be covered with polymer-modified bitumen was the 600-metre St Quentin bridge in France, in 1976. Samples of the surface were removed for analysis in 1985, and tests showed that the binder had not undergone any significant changes. A particularly challenging road is that through the Mojave Desert in the United States; temperatures can vary between a day-time high of 42 °C or more, and a night-time low of 0 °C, and it is traversed by a large number of heavy commercial vehicles. Tests done by the Californian Department of Transport on the desert Interstate Highway 40, in which modified bitumen was laid side by side with ordinary asphalt, proved conclusively that the new form is much better. The ordinary asphalt showed signs of wear after two years and serious cracking and fatigue after four, while the modified bitumen was free from such defects.

Modern roads are also much quieter when they are laid using porous asphalt (which is also known as 'whispering asphalt' because vehicles driving on it produce much less noise, or 'popcorn asphalt' because of its texture). It provides four benefits: it cuts traffic noise by half; it reduces spray when the road is wet; it minimizes glare in wet night-driving conditions; and it saves fuel by reducing the friction between tyres and the road surface. Porous asphalt reduces the traffic noise that comes from compression and expansion of the air trapped in the tread of the tyres; this type of noise very much depends on the texture of the road surface. If the road surface is hard, it may also increase traffic noise by reflecting the sounds that come from the underside of the car, from the gearbox housing, and the engine.

Porous asphalt is made with so-called 'gap-graded' aggregates, which are a type of aggregate that has been partly graded to ensure that it will not pack very effectively. The main difference between ordinary asphalt and porous asphalt is the percentage of empty space. Ordinary asphalt has only about 5% of its volume composed of voids, whereas porous asphalt has at least 20%, and generally more. There are three types of aggregate, depending on its size: continuously graded, gap graded, and uniformly graded. As its name implies, continuously graded aggregate has pieces of all sizes, and as such it packs well, leaving few voids. At the other extreme is uniformly graded aggregate, with pieces that fall within a particular size range. Such aggregate does not pack effectively, and leaves many voids. Intermediate between the two is gap-graded aggregate with a higher percentage of larger pieces, also ensuring that there will be a number of voids.

Porous asphalt made using unmodified bitumen has one major drawback: it does not last as long as ordinary asphalt. There may be more than one reason for this. Its open structure allows more contact between the bitumen and oxygen and water, leading to a more rapid ageing. At the same time, there is less support to enable it to resist the forces on the road, and this leads to the loss of stones from the surface. Its pores can also become clogged with soil, thereby reducing its effectiveness. It can then become prone to frost damage if the water cannot drain away and freezes. Bitumen modified with SBS reduces the rate at which these damaging processes occur.

Roofs have also improved, thanks to better bitumen. Roofing felt, once seen as the cheap and quick way to cover industrial buildings, used to be made from a layer of glass wool or polyester mat sandwiched between two layers of bitumen-based material, about 1.5 mm thick, which contained powdered limestone or talc. Following the oil crisis of 1973, when cheap heating fuels were no longer available, roofs began to be insulated as a means of conserving energy. This raised the temperature of the roofing felt because it meant that heat from the sun could not be dispersed into the building, so the felt got hotter during the day and colder at night. This increased temperature fluctuation made it liable to develop cracks and thence to leak.

The answer was to modify the bitumen by adding SBS polymer, and as much as 12% of this is incorporated into the bitumen by mixing the two at 180 °C. The resulting bitumen is highly elastic— it can be stretched to more than fifteen times its own length before it breaks—and it can even be coloured. As a result, this new type of roofing felt is being used on offices and public buildings, and even for churches.

Orimulsion

As everyone knows, oil and water do not mix, and the same is true of bitumen and water. But oil and water can be made to *appear* to be mixed, as an emulsion, if the droplets of oil are made small enough. We see this in such products as homogenized milk or mayonnaise. Again, the same is true of bitumen, and the resulting fluid is called orimulsion. It comes from the Orinoco delta in Venezuela, where there are an estimated 1200 billion barrels of bitumen in what is the largest untapped reservoir of fossil fuel on the planet.

Orimulsion is an emulsion of bitumen in water, and is kept in this state by the addition of small amounts of the surfactant nonylphenol ethoxylate. It was developed in the 1980s as a result of joint research by Petroleos de Venezuela, the state-owned oil company of Venezuela, and the oil giant BP International. The surfactant keeps the bitumen suspended in the water in the same way that the surfactants in detergents keep grease suspended in water, by coating each droplet. The droplets of bitumen are about 20 microns in size (20 millionths of a metre), and they make up 70% of the total volume. The result is a fluid with the viscosity of medium-grade crude oil, and like the oil it can be pumped along pipelines or shipped to where it is used. It burns more efficiently than normal fuel oils, releasing 99.9% of the energy it contains, and, because of its water content it burns at a lower temperature, which serves to cut the emission of some pollutants, such as nitrogen oxides and particulates.

Not everyone considers orimulsion to be a useful resource. Indeed, whenever orimulsion is suggested as a possible fuel, it provokes opposition from environmentalists who label it the 'fuel from hell', generally because of its high sulfur and metals content. (And in this respect they are justified, because it contains nearly 3% sulfur, 100 p.p.m. of nickel, and 400 p.p.m. of vanadium.) Orimulsion has been used as a fuel for power plants and cement works, and has been shipped to the United States, Britain, Germany, Italy, Denmark, Japan, and Canada, where its use has sometimes met with vociferous opposition, often sufficient to get it banned. While it is now much less used in most of these countries, however, it is still shipped to China, Korea, Philippines, and other South American countries. More than 6 million tonnes of orimulsion were exported from Venezuela in 2001.

Polycarbonate

Polycarbonate is the polymer of the Information Age, most visibly used to make CDs and DVDs, mobile phones, and global satellite monitors. The reason for its success lies with its incredible strength, durability, and colourability, and the fact that it can be blended with other polymers. Yet there are those who doubt its safety.

People who disapprove of disposable nappies may also disapprove of polycarbonate baby bottles. Indeed, it seemed as though they might have good reason for doing so as a result of an article entitled 'Baby Alert' in the US magazine *Consumer Report* in May 1999. This was followed by a documentary on ABC television, and the article and the documentary warned of the dangers of bisphenol A (BPA). Traces of this chemical were present in the polycarbonate plastic of which feeding bottles were made, and there was evidence that it could leach into the bottle's contents, and might well affect the sexual development of the child because bisphenol A was said to be a 'gender bender'.

Polycarbonate was first made by a German chemist, Gunther Einhorn, in 1898, who reported that it formed as an insoluble solid in the reaction vessel in which he was trying to make organic carbonates. Two other chemists, Bischoll and Hedenstrom, produced more of it in 1902, and, while it was an interesting material, it appeared to be of little use. That was also the opinion of the great polymer chemist Wallace Carothers, the man who worked for the chemical company Du Pont and who discovered synthetic rubber and nylon in the 1930s. Indeed, it was not until 1953 that the Bayer laboratories in Germany were able to produce a sample of polycarbonate that was suitable for commercial applications. It came on to the market in the 1960s under names such as Xantar, made by the Dutch chemical company DSM, and Lexan, made by General Electric in the United States. Today it is manufactured by the chemical companies Bayer and Dow, and they are currently jostling to oust General Electric from its number-one position as the world's largest polycarbonate producer.

Polycarbonate is strong, hard wearing, and tough, and it remains stiff up to 140 °C and resilient down to −20 °C. It does not easily burn, and with the addition of flame retardants it will pass severe flammability tests. It is not affected by most chemicals. Polycarbonate will blend with other polymers, and the blend with ABS (acrylonitrile butadiene styrene) is even stronger than polycarbonate by itself. Polycarbonate is naturally transparent, but it can easily be coloured and its surface embossed.

Polycarbonate is manufactured from BPA and carbonyl chloride ($COCl_2$, also known as phosgene). BPA has various uses, but two-thirds goes into manufacturing polycarbonate. (The rest goes into making resins and flame retardants.) BPA is made on the 50,000-tonne scale in plants in Europe, the Far East, and the United States.

Bisphenol A consists of two phenols connected to the same carbon, which also has two methyl groups (CH_3) attached. When BPA reacts with carbonyl chloride, in a solvent, it polymerizes to form chains in which carbonate units (CO_3) interconnect BPAs. More than 99.9% of BPA never leaves the site where it is manufactured, but is immediately converted to other products. Even the small amount of vapour that is lost poses no threat because it is rapidly degraded by sunlight, and any that gets into water or soil is biodegraded within a day or two. Despite there being seemingly little risk to humans or the environment, there were concerns expressed about BPA once it had been shown to have endocrine activity.

'Gender benders' is the environmentalist phrase for endocrine disrupters, or more specifically for those that behave like oestrogen, the female hormone. Endocrine disrupters are substances that can upset the action of hormones, and not just those concerned with sexual development, although this is what is generally assumed to be the threat they pose. Other chemicals, such as the pesticides lindane and tributyl tin, have been shown to act this way. There are several manufactured materials like these which, in the past, were used but have been banned, and rightly so. There are some, however, for which there is no evidence that they disrupt our hormone systems, or even those of other animals, and BPA is one such. Yet such was the clamour against it that the US National Institute of Environmental Health Science, and the US National Toxicology Program, set up a joint committee to look into the matter, although they eventually concluded that BPA appeared not to have any low-dose effect.

There is no denying that BPA can act as an endocrine disrupter; this was discovered by accident at the Stanford University School of Medicine in Palo Alto, California, where research into yeasts was being hampered by an unknown contaminant, which was eventually traced to BPA leaching from the polycarbonate flasks in which the research was being conducted. Although the concentration of BPA was only 5 p.p.b., it was enough to cause an oestrogenic response in the yeast cells. BPA has only a weak oestrogenic action and in no way compares with the effect of oestrogen itself, being thousands of times weaker. Nevertheless, that it could exert an oestrogenic effect was enough to condemn it in the eyes of some, such as Theo Colborn, Dianne Dumanoski, and John Myers, authors of the book *Our Stolen Future*. The World Wildlife Fund for Nature also has reservations about BPA.

Environmentalists claimed there were detectable levels of BPA in natural waters near manufacturing plants, and this, too, led to public concern. The levels measured were less than 1 p.p.b., and in some rivers supposedly at risk no trace of BPA could be found. In any case, the no-observed-effect-concentration (NOEC) for a lifetime exposure to BPA for daphnia, the most sensitive freshwater organism, showed that 1000 p.p.b. had no effect. Indeed, there have been no reports of BPA affecting any aquatic system.

What also drew attention to BPA as a possible endocrine disrupter were reports from Frederick Vom Saal and co-workers at the University of Missouri in Colombia. They said that a small number of mice fed low doses of BPA were found to have reproductive disorders. According to Vom Saal, these baby mice then suffered long-term effects: females reached puberty earlier and gained weight more quickly than normal. The use of epoxy resins in dentistry came under scrutiny when Vom Saal said that, in the hour after a filling had been done, a 'significant' dose of BPA leaked from it. Polycarbonate was used as a dental sealant, and clearly this was a worrying finding, but other tests showed that there was no leakage of BPA, and the American Dental Association has confirmed this to be so.

A minute amount of BPA is indeed released from the surface of polycarbonate into any liquid contained in it, and this may be as little as 1 p.p.b. Even if it were several times this level—and others claim to have found 5 p.p.b.—it is still tiny. (Five parts per billion is the same numerically as 1 *second* in 6 years.) If a baby is fed on formula feed with this level of BPA, then it would be taking in about a microgram of material a day, which is far less than the level at which any physical effect would be observed and about fifty times lower than the amount regarded as a 'safe' daily dose. In fact, *Consumer Reports* appeared to have its sums wrong and was quoting levels many times higher than that which it regarded as the safe dose.

Meanwhile, the manufacturers of BPA say that there is no evidence that this chemical poses any risk to human health and that it has been thoroughly tested according to procedures laid down by the US National Academy of Sciences. BPA is not carcinogenic or mutagenic, and human exposure is several hundred times less than that at which any effect is likely to be observed. Those who come into contact with things made from polycarbonate are unlikely to imbibe any BPA, considering the nature of the things that are manufactured from it: CDs, DVDs, riot shields, vandal-proof bus shelters, cooling fans, aeroplane windows, bullet-proof glass, protective canopies, skylights, conservatory roofs, safety helmets, headlamp covers, lighting, mobile phones, battery boxes, household appliances, dashboards, cash dispensers, global satellite locators, and car body panels and bumpers. World demand for BPA is growing faster than for any other plastic, and production now exceeds 2 million tonnes a year.

The only uses of polycarbonate that might put traces of BPA in the human body are baby-feeding bottles, packaging, water dispensers, and medical apparatus. Such uses were given a clean bill of health by the US Food and Drugs Administration, and the minute amount of unreacted BPA that might migrate into the food was considered to be safe. Nor did BPA pose an environmental threat, and the Organization of Economic Cooperation and Development (OECD) classed it as readily and totally biodegradable.

But it was the research that appeared to have shown that BPA was a gender bender that alarmed the media and could not be ignored. Two industry bodies, the Society of the Plastics Industry and the European Chemical Industry Council, sponsored a large-scale investigation. Research was undertaken in various laboratories, using larger numbers of female and male rats and mice, exposing them to a wider range of BPA doses over longer periods of time. The researchers could find no effects on prostate, sperm counts, and testicles in the males; for the females, there were no bodily changes in their growth and development, their ability to conceive and deliver, or in their offspring and their offspring's growth and development. When BPA was given to pregnant rodents at levels that produced a toxic reaction, they still gave birth to normal babies.

One of the conclusions of all this research activity was that it is very easy to confuse other chemicals with BPA. In fact, no BPA was detectable in the contents of beers and drinks sold in cans lined with resin, at least down to a level of 5 p.p.b., the limit of detection. Other canned food did show levels of BPA, at around 35 p.p.b., much less than had previously been claimed; food from such cans would provide a daily intake of BPA but it would be 500 times lower than the safety level suggested by the FDA.

The Society of the Plastics Industry was relieved by these findings, saying that none of the effects the University of Missouri researchers claimed to have observed could be replicated. The debate about the testing of BPA as an endocrine disrupter led to an angry exchange of letters between industrial and academic investigators, each side accusing the other of misleading statements based on poorly designed experiments giving ambiguous results. So who is right? One's sympathies lie with Vom Saal: because he is an academic, one assumes his motives and objectives are above reproach. Yet the researchers employed by the manufacturers were no doubt equally qualified, perhaps even better resourced and more experienced, and some were clearly working independently.

This placed officials in a somewhat difficult position over whom to believe. At the request of the Environmental Protection Agency, the US National Toxicology Program issued a preliminary report in May 2001 pointing out that the inconclusive findings thus far obtained meant that more research was needed, and until that happened that the current EPA guidelines should remain in force.

Rochelle Tyl was responsible for a major study into bisphenol A that was carried out by the Research Triangle Institute, North Carolina, in the United States. This studied the effects of high and low doses of BPA on three successive generations of laboratory animals, and came to the reassuring conclusion that it has no effect on the reproductive system. The work was sponsored by the Bisphenol A Global Industry Group, and was given approval by the US National Institute of Environmental Health Science, the National Toxicology Program, and the Environmental Protection Agency. That was in the year 2000. Then, in 2003, came a report from Patricia Hunt, of Case Western Reserve University, who had found that mouse eggs exposed to low concentrations of BPA developed abnormalities. She had discovered that polycarbonate vessels which were washed with a solution of alkali detergent thereafter leached small amounts of BPA, enough to cause chromosomal abnormalities.

Keeping an eye on the future

There is a second type of polycarbonate that is very different from the one based on BPA. It is made from a starting material that has two carbonate groups linked by CH_2CH_2 and two double-bond-containing groupings at either side. It is these that undergo **free-radical polymerization** (see Glossary), the one to form long chains, the other to link these chains together. This cross-linking makes the polycarbonate particularly strong, and it is ideal for the lenses of spectacles because it has a higher refractive index than glass. This means that the lenses can be thinner and ultra-light, and are ideal for those who used to have to wear thick lenses, and who then looked like the archetypal 'goggle-eyed' scientist.

In this respect the new polycarbonate may well help create a fresh image for chemistry, and so help attract bright young people to study and research it. Indeed, they will be sorely needed in this century if the world is to base its economic well-being entirely on renewable resources.

Chemiphobia, its Causes and Cure

I AM A CHEMIST and naturally supportive of the benefits that chemistry has wrought in removing the fear of famine, disease, and poverty from the lives of most who live in countries where a strong chemical industry has developed. It would be reassuring to hear the occasional expression of gratitude for what has been achieved, but when people in the United Kingdom were asked whom they most trusted, they put those of the chemical industry in the bottom category along with used-car salesmen. I'm sure the same would be true in other European countries and North America. Of course, Shakespeare knew that such a reaction is typical of human beings:

> AMIENS (*sings*) Blow, blow, thou winter wind,
> Thou art not so unkind
> As man's ingratitude. . .
>
> Freeze, freeze, thou bitter sky,
> That dost not bite so nigh
> As benefits forgot.
>
> [*As You Like It*, Act II, Scene vii]

To some extent the industry itself brought much of its current poor image upon itself, by polluting the environment and being inadequate in its communication skills. While the first of these is no longer a problem —the waste from agriculture, mining, and metalworking is far worse, and always has been—the second remains. Chemicals are endlessly pilloried by various groups that oppose the chemical industry, and they seem to be able to deliver their negative messages and alarms with impunity, or at best with only weak rebuttals. In this Postscript I shall try to analyse how they do it, and what safeguards we might use to ensure that they don't end up killing a goose that continues to lay some of the golden eggs I have been writing about.

Why is it still possible to alarm people about minuscule traces of contaminants in the water they drink, the food they eat, and the air they breathe? People in the West have never been healthier or lived longer. Here, I am talking about conditions in developed countries, where the chemical industry has been established for a long time and public health has been on the agenda for more than a century. Compared with the obnoxious materials in the water, food, and air of a century ago, today's environment is clean and safe, and getting better all the time.

So how was it that, in the 1980s and 1990s, the media endlessly retailed a stream of alarms about threats from 'chemicals'? I put the word in quotation marks to show that the meaning of this word is not one I would choose. Most people today take 'chemical' to mean toxic or polluting, and something to be avoided as dangerous almost by definition. As a chemist, I use the word chemical non-judgementally as meaning simply a substance, although I now tend to avoid using the word wherever possible because of its negative connotations.

At this point, I am sure that many readers will be thinking: but surely there must be some truth in all these reports of the dangers of chemicals, otherwise why would we hear so much about them? Were not the things that we have been told based on *scientific* evidence? The answer is often no. The science behind the oft-used phrase of 'scientific findings' was just not there. The evidence that condemns various chemicals is the end product of three activities that are carried out in a rather obscure way: the collecting of the data, the manipulating of the data, and the publishing of the data. It's time to look at how we can be misled by this sequence of events.

Epidemiology is a respected branch of learning, traditionally based on gathering data about infectious diseases that affect large numbers of people and looking for clues as to the underlying cause. (The world 'epidemiology' derives from the word 'epidemic'.) Today, it is still focused on illness, but is now concerned with non-infectious disorders such as cancer and heart disease, and the occurrences of such conditions are charted with a view to linking them to other factors that might give some insight into their cause. Often those supposed factors are 'chemicals',

and especially the chemicals that are in the products we   purchase or in
our immediate environment.

Collecting the data is one activity, processing them mathematically is
another. Faults in either can render the results worthless. Professional
epidemiologists are aware that, if findings are to mean anything, then
care has to be taken to avoid the many pitfalls that await the unwary, and
they have refined their methods to ensure these traps are avoided in both
halves of the process. Many other scientists who use epidemiological
methods almost always fall into one of them—not that this prevents
them publishing their results and claiming them to be scientifically
based. The findings that reach the media may, in fact, be based on no real
science at all, and they often come from a special-interest group with
popular support.

The first pitfall is that the group already has a point of view they believe
is true and want to prove. Let me take an example of how it might be
done. Suppose you believe that the increase in asthma cases is caused by
the increasing use of air fresheners in the home. You start by asking
those who have asthma if they have ever bought one of them. You then
ask the same question of a similar number of people who don't have
asthma. If, after interviewing 100 people in each category, you find that
50 of those with asthma say they have breathed artificially fragranced air,
but only 45 of those who don't have asthma have been so exposed, then
you can announce that there is a link between this condition and air
fresheners, and that there is a worrying 12% increase[1] in asthma among
those who use them. You have 'proved' what you set out to prove, that
here is yet another instance of 'chemicals' causing disease. Of course, no
reputable scientific journal would publish such findings, but you might
get them into print in another way.

Write a press release to announce these results and, provided you are
working for a respected institution such as a university, you are almost
certain to make the headlines, although these might be a little exagger-
ated: 'Air-freshener asthma link' or 'Fragrant route to asthma'— you
know the sort of thing. You might then apply to some funding agency for
a large grant to do essential follow-up research into what is clearly an

---

[1] The increase difference is 5, which is expressed as a fraction of the lower number, 40.

issue that is alarming the public. Of course, such barefaced manipulation of the media should never occur, but it sometimes appears to be the explanation.

Clearly, any epidemiological study has to involve a *significant* number of people, and these have to be in two groups: those who have the condition being investigated and those who do not. Obviously, the two groups have to be matched in terms of age, sex, weight, marital status, ethnic group, social class, occupation, and general behaviour (such as smoking and drinking), although this might prove difficult to achieve if you have limited resources.

Even if you meet these conditions, there are still pitfalls to avoid. These were outlined very clearly by Göran Pershagen of the Department of Epidemiology of the Institute of Environmental Medicine at the Karolinska Institute in Stockholm, Sweden, in the book *What Risk?* Pershagen identifies three kinds of epidemiological error: selection of the data, mistakes with the data, and hidden factors influencing the data. Each of these errors can arise in more than one way. For example, data selection may be skewed towards the desired result by having a sample of people that is not chosen randomly from among the population, or by deciding that some of the data are 'wrong' and so can be omitted. When that happens, the tendency is to exclude data that are not in agreement with the expected outcome. The selection of data may even be done unconsciously, which is why it is so important for it to be done by a totally independent survey organization. This, of course, is rarely the case because of the expense involved.

Mistakes with the data can arise in a variety of ways, but the most obvious one is having to rely on those being interviewed telling you the truth. Sadly, this is often not the case and was shown quite clearly in an epidemiological study on the effects of diet. When families were asked about their eating habits, and then the contents of their refuse sacks were analysed—without their knowledge—there was little correlation between what they said they ate and what they actually ate. Even when people tell what they perceive as the truth, they may in fact be selective in their answers. For example, the question 'How many times a week do you have sex?' generates a reply that more nearly relates to what they do during the two or three weeks when they are on holiday, rather than what they do at home.

The third type of mistake is the hidden factor, referred to by epidemiologists as the confounding variable. This can throw doubt on otherwise well-constructed epidemiological studies, as it did with those that linked drinking red wine with lowered risk of heart disease. There the hidden factor was social bias, and this was revealed in 1997 at the Novartis Foundation Symposium on Alcohol and Cardiovascular Disease, held in London. It was pointed out that those who drink red wine tend to be middle class, and they as a group live longer anyway, while those who rarely drink red wine tend to be the poor whose average life-span is shorter.

Unless an epidemiological survey has been carried out under the watchful eye of a trained epidemiologist, then it is unlikely to be of much value, but that does not deter those who wish to prove a point. There are several little tricks that they can, and do, pull when presenting their results. Most of these come under the label of 'statistics'. Again, there are trained statisticians who can process large amounts of data in perfectly sensible ways and often draw out from them connections that at first sight are not obvious. They are well versed in deducing the reliability of the conclusions that they make, and surround these with assessments of how reliable they are. Amateur statisticians, however, are not to be trusted.

'Statistically significant' is a phrase that seems to assure us that the correlation is, in fact, based on fact. A useful rule is that for a conclusion to be worth considering, we need to be assured that there is a 95% possibility that it has not been arrived at by chance. (This still means that one conclusion in twenty will be simply wrong, and these are not insignificant odds, but this reflects the level of uncertainty that we live with everyday, and so it is acceptable.)

So what are the tell-tale signs that should alert us to 'findings' that are mere artefacts of the way the data have been collected or the way they are being presented?

One warning sign is to see findings that are presented as a graph with no origin. The origin is the 0,0 point, at which there is no input of the substance being tested and so no effect on the population being studied. Graphs without origins are useful propaganda ploys for financiers who want to show the recent performance of their investment products, and politicians who want to show improvements in their standing or the effects their policies are having. In epidemiological surveys, the omission of the origin is the only way to disguise the fact that some level of exposure to the investigated substance might have no effect at all, and that an effect becomes apparent only after a certain level is reached.

Those who hide the origin are trying to disguise the fact that low levels of exposure may pose no risk, although we know it may be dangerous in larger doses. The usual argument is to assume that because it is dangerous at some level it cannot be safe at *any* level, and that it is intrinsically hazardous per se. The toxic effect of selenium proves that this kind of assumption is basically flawed, as we saw in Chapter 3. The dose really does make the poison. In fact, the human body has several in-built safety mechanisms for removing unwanted components in our food, such as natural toxins. All plants contain potent pesticides to protect them from microbial attack and insect predators, and we eat these along with the parts of the plant that provide us with the nutrients we need. In fact, we have quite a high threshold against such chemicals having any effect on us.

Epidemiological and statistical data should never be accepted as reliable unless they come with the margin of error. This plus-and-minus ($\pm$) number tells you the upper and lower limits of the data point. The more people that take part in a survey, the smaller the margin of error should be. If you are being quoted percentage data, then the margin of error can easily be assessed, provided you are told the number of people on which they are based. Any survey that does not reveal this number should be dismissed out of hand.

The margin of error is calculated as 100 divided by the square root of the number of people surveyed. If a thousand people have been interviewed and the pollsters say that 50% are in favour of a certain proposal, then the true number probably lies between 47% and 53%, because the square root of 1000 is 31.6, and 100 divided by 31.6 is 3% to the nearest whole number. Thus, the average of 50% has a confidence limit of between 50 − 3% and 50 + 3%, that is, between 47 and 53%.

If the sample size were only 100 people, then the margin of error would widen considerably, to $\pm$10%, and if the researcher had only 50 responses to analyse, then the range around a result reporting a figure of 50% would be $\pm$14%, and virtually worthless because it admits that the real figure is somewhere between 38% and 64%. Yet some data have perforce to be based on such small numbers. Moreover, by the time the data hit the media, only the upper figure is emphasized, although it might be qualified with a phrase such as 'as many as 64% of people could be affected by . . .'

When reading about percentage differences, bear in mind that percentages calculated on fewer than 100 are highly suspect. The idea of percentages is to express very large numbers on a scale that we can more easily comprehend. Examples would be that 5% of the population is unemployed, or that 20% of the national budget is spent on defence. It is pointless converting a small number into a percentage because you are levering it upwards unnecessarily, although if your 'findings' are small this is one way to make them seem important. For example, if there are eight cases of a rare cancer each year in city X where there is an incinerator that might be emitting dangerous fumes, and only six among the population of city Y where there is no such incinerator, you might not find this particularly alarming. If, however, you say that there is a 33% greater risk[2] of contracting the cancer in city X, without giving the actual numbers on which the data are based, then you can make your point in a way that will grab the headlines. Needless to say, this kind of survey is so riddled with confounding variables as to be worthless, except for publicity purposes.

The press release

It is the job of newspaper sub-editors to devise headlines, and they know that readers' attention can best be caught with a carefully baited emotional hook. Pressure groups who seek to capture public notice are skilled in providing such hooks, often to publicize 'findings' which they themselves have generated (a) by directly commissioning a survey (where they influence the questions to be asked), or (b) indirectly, by funding research from experimenters they know to be supporters of their cause.

These are the sources of many of the scares that involve chemicals, and there is an endless flow of them. They come from several pressure groups: environmentalists blame chemicals for causing pollution; health gurus blame them for causing cancer; and organic food producers say they ruin the soil and contaminate crops. These pressure groups are skilled in the art of writing eye-catching press releases, and naturally hard-pressed journalists find them appealing. Because they come from groups that have widespread public support, the journalists assume that

---

[2] A calculation based on the number in a city Y. The increase is $(2 \div 6) \times 100 = 33\%$.

the findings are trustworthy. Sadly, often they are not. The aluminium story in Chapter 5 shows how badly they can get it wrong.

By itself, an argument which appeals to reason should convince you, but advertisers have long known that to carry a message to the general public you have to use strong emotions if it is even to be noticed, let alone become lodged in the national psyche. The same is true for science, whether we want it to be that way or not, but scientists rarely resort to emotional 'hooks' to catch media and public attention. People may read that a new drug is a major step forward in treating a particular disease, but the information will be quickly forgotten unless the news causes a visceral reaction by evoking feelings such as fear for oneself, or sympathy for others.

The emotional hooks most used by those who find chemicals threatening, and who wish to show that they bring only trouble, are of three kinds: those that relate to individuals and their families; those that concern us as members of a wider community; and those that suggest the larger world is being put at risk.

A common ploy of the first type is to suggest a threat to a highly vulnerable group towards whom we are naturally sympathetic, such as babies, breast-feeding mothers, and young children. Threats to fertility also play on the fears of men that they may become infertile, or of women that they may give birth to a deformed baby. At the other end of life's journey is the worry of dying, and naturally people are concerned if they read of things causing heart disease or cancer. Finally, there are the warnings of global disaster through threats to biodiversity, or changes to the atmosphere or oceans. The menace-loaded phrases of the alarmists have now entered the language as 'acid rain', 'toxic chemicals', 'gender-benders', 'Frankenstein foods', 'cancer on tap', and 'artificial fertilizers'.

Carefully orchestrated campaigns can persuade people to adopt attitudes that common sense should tell us can only lead to disaster. For example, the organic movement would like to see an end to all agrochemical fertilizers and pesticides, but that would mean converting almost all of the Earth's fertile land to human food production if millions were not to die of starvation. These chemicals have made it possible to grow *four* times as much food on a farm as can be achieved by organic farming methods. And it has environmental benefits that are rarely mentioned. Thus, if 1 hectare of land can support twenty people, instead of just five, then 3 other hectares of land can be left undisturbed for us, and wildlife, to enjoy.

To sum up: when confronted with news items clearly based on a press release, you should ignore any 'findings' that include the following:

1. Those presented as a graph, unless it includes the origin.
2. Those that do not include the margin of error.
3. Those based on only a small sample size, or a the sample whose size is not revealed.
4. Those that are presented only as percentages.
5. Those that play on our emotions.
6. Those that use the language of uncertainty, with words and phrases like 'could be linked to', 'is believed that', 'appear to show', or that quote unspecified groups, as in 'doctors are convinced that', 'scientists now believe', etc.

All the foregoing comments are simply a warning to examine more closely the opinions of those who seem well intentioned, but who may really be covert chemiphobics. Their successful campaigns against chemicals would not have been thought creditable by previous generations, but they have become more effective and influential, even to the extent of changing the law in some countries. Indeed, in Europe it will soon be mandatory to test all chemicals for safety, including those that have been safely used for a hundred years or so. The cost will not only be in monetary terms, and the lives of millions of laboratory animals, but may also cost Europe its chemical industry, with the loss of millions of jobs.

In an age when feelings are seen as more important than rational thought, it is easy to be swept along by emotion, but while it is possible to defy reason with emotion, in the end even emotions cannot defy the laws of chemistry. In *Vanity, Vitality, and Virility* I have tried to be as objective as possible when looking at the thirty or so chemicals that you have been reading about, pointing out their drawbacks as well as their benefits, but I am aware that some readers will have wondered if these chemicals of everyday life really are safe. You may even have suspected that they could be affecting your health, perhaps with some long-term effect that we don't yet appreciate. Some of you may well wonder whether we should always adopt the so-called Precautionary Principle: in other words, if there is any doubt about a new chemical, let us not introduce it until we can be sure that it will have no harmful effect whatsoever.

Most of us live in urban areas where we breathe air that contains some noxious fumes; we drink tap water that we have been told could be less than healthy; and we eat processed food that is dismissed by some as unhealthy on account of the additives it contains or the way it has been processed. There is a little truth, but only a little, in all these allegations. Yet there are those who believe that everything is now massively polluted with unnatural chemicals, and the depth of their concern can be judged by the lengths to which they go to avoid them. I hope that *Vanity, Vitality, and Virility* will have set your mind at rest with respect to some of them, or at least shown that the threats they pose are worth the risk because of the benefits they deliver.

In the last quarter of the twentieth century, there were many books that suggested that food additives were a major cause of human illness, especially cancers. Reassurances that these substances had been tested (and found to be safe) were countered with suggestions that 'long-term' effects could not be so assessed, and surely it was better to ban them altogether. When it was pointed out that traces of natural toxins in foods are proven carcinogens and that these could not be banned, these arguments were waved aside on the grounds that human exposure to them had given us some kind of protection, presumably on the basis that those who were susceptible had been weeded out by a million years of evolution. It sort of makes sense.

Nevertheless, what is sauce for the goose is sauce for the gander, and it was not long before natural remedies were being tested in the same way as the hated chemical additives. As a result, herbs like comfrey were banned, and health warnings issued against many traditional remedies and Chinese medicines, not to mention the dangers of drinking too much coffee, or of eating barbecued meat and fried foods.

Those with a mischievous turn of mind might even try turning the argument on its head. Suppose the chemicals that we were adding for one purpose, possibly merely to add colour to food, were actively *preventing* disease, perhaps by boosting our immune system or even attacking invading pathogens. That seems about as likely (or unlikely) as the thesis that they are weakening our immune system and/or causing disease.

We've no proof either way, you might think. When tests have been carried out on some food additives, however, the rodents they were tested on were found to have *fewer* cancers. (That discovery did not make the headlines, of course.) Consequently, it may well be that *trans* fats, like CLA, really can protect against breast cancer, as was suggested in Chapter 2.

Of course, once something has been accused of being a threat, it is almost impossible to prove it is safe. Proving a negative is the equivalent to being accused of a crime and then you, the accused, having to prove your innocence. In all civilized societies, it is up to the accuser to prove your guilt, but when emotion rules, as it did in the time of the witchcraft trials, then the reverse attitude prevails.

So what should we do? Surely we must stop finding a chemical guilty merely because someone has accused it, and at least ask the *accuser* to provide evidence of its guilt. Speculation alone is not enough, although it has condemned many useful substances in the past. The traditional test of logical deduction based on meaningful scientific evidence is the minimum we should expect before we launch into any campaign to ban a chemical. Who knows, we might already have banned a chemical whose 'long-term' effect might have been to protect us against Alzheimer's disease in old age. This is most unlikely, but logically it is no more unlikely than assuming that chemicals that have been rigorously tested and safely used for decades must contain some hidden danger that we don't yet know about.

This Postscript is really my attempt to win over those chemisceptics among you, and perhaps even give the chemiphobics pause for thought. They see all chemicals as threats, and must therefore live in an insecure world of hidden fears, perhaps even imagining that a single unnatural molecule might be all that is needed to trigger cancer or heart disease. For them, and there are probably many who now think this way, the daily drizzle of chemicals in food, water, and air is the cause of all their ills.

For thirty years or so, there have been developments that have undermined the scientific approach to knowledge: the growth of the media and their thirst for news, which, of course, is no bad thing, but it has become a threat when used by special-interest groups and especially those movements opposed to the chemical industry. The result has been a radical shift in the public's perception of the benefits of chemistry. Perhaps all sciences that are successful and move ahead too quickly are bound to provoke such a reaction, because the same has happened with the nuclear industry, and currently with the biotechnology industries. It has yet to befall the electronics/communications industries, although the Orwellian possibility of monitoring everyone's movements all day and every day may soon lead to a reaction there.

I would like to end the book with a few general points about chemicals, to reassure you that your health is not being endlessly undermined by the products of the chemical industry.

Cancer is a risk, but the risk may be so low as to be worth ignoring.
Chemicals that cause cancer in specially bred laboratory rats are unlikely to cause cancer in humans. Many of the chemotherapy drugs used to treat cancer would be classed as 'cancer-causing chemicals' on the basis of such tests, but are used because they are so good at attacking primary cancers that the tiny risk of them causing a new one is unimportant if a life is to be saved.

There is no such thing as the 'cocktail effect', as applied to chemicals.
Chemiphobics believe that, in the wider environment, chemicals have a synergistic effect on one another: in other words, they boost their respective effects. This is often referred to as the 'cocktail effect' by analogy with the belief that mixing alcoholic drinks makes them much more potent. (Even for alcohol this belief is unfounded.) The cocktail effect can happen, and we know that artificial sweeteners together produce a stronger effect than when used singly, and the same is true of painkillers. But what is true when we are talking about related chemicals acting on a group of receptors, be they on the tongue or in the brain, does not apply to unrelated chemicals in the environment, especially when they have been tested and no such 'cocktail effect' has been registered. Environmental chemicals may add up to exert an effect, but they do not *multiply* that effect.

Synthetic chemicals are rather feeble endocrine disrupters compared with the hormone mimics that plants produce.
Of course, there are some drugs that are meant to act this way—the contraceptive pill being the best-known one—but these are specifically designed for the job. Endocrine-disrupting chemicals (oestrogen mimics

and oestrogen antagonists) are naturally abundant in the environment (soya producing one of the more potent ones). They are produced on a large scale by plants to protect themselves against insect predators,but they have very little effect on humans. Thus, natural endocrine disrupters are far more potent than synthetic environmental chemicals. For example, a glass of red wine contains 99.99% of the endocrine disrupters consumed in a three-course meal, and it is still not clear that the remaining 0.01% is the result of pesticide residues in the rest of the meal. This small percentage, too, might be due to natural chemicals.

We come into this world with a wonderful built-in detoxifying system that protects us against thousands of natural toxins every day.
Human metabolism has evolved so that we can detoxify unwanted and potentially harmful chemicals in our food supply, provided we do not overload our body with such chemicals. Living cells have also developed ways of rapidly repairing any damage caused. When talking about chemicals, however, the common assumption is that no dose, however tiny, is without some harmful effect. This view is wrong and we now know that there is a threshold level that our excretory mechanisms can deal with safely.

Current analytical methods enable us to detect extremely low concentrations of chemicals.
Some chemicals are present in such tiny amounts that they have to be measured in parts per trillion (p.p.t.), which is like one *second* in 30,000 years. Those measured in parts per billion (p.p.b.) are still very small: in this case, 1 p.p.b. is like one second in 30 years. Even one part per million (1 p.p.m.) is still only like one second in 12 days. Nevertheless, it is still possible to make even these small amounts seem menacing, so that instead of saying that chemical Z has been measured in breast milk at only 0.1 p.p.m., I would, if I were trying to prove it was a hazardous threat, say it is there at an alarming 100 p.p.b.

Finally, a word to any young person reading this book who might now be contemplating a career in chemistry and the chemical industry: join us. Chemistry is an exciting science, and becoming even more so as molecular biology and nanotechnology fall within its remit. In *Vanity, Vitality, and Virility* I have tried to show that chemists can work wonders, although we still can't work miracles. The biggest challenges that the world faces this century are to continue to enjoy all the products of the chemical industry, yet produce them from renewable resources, and to spread their benefits to all the world's people, not just those in developed societies. Chemistry can lift the threat of disease, malnutrition, and poverty from everyone on this planet, and yet not plunder the Earth to do so. It can be done, but it will require bright and committed young people to do it. Chemistry is a hard science, and those who choose it as a profession deserve our admiration, and not the unjustified suspicion that has driven so many away from studying it in the past 25 years. We lost one generation of support, but the world cannot afford to lose another, or else a new Dark Age will really be about to dawn.

Glossary

Alpha-hydroxy acid (AHA). This describes a molecular arrangement in which a carbon atom has attached to itself both a carboxylic acid group (CO_2H) and a hydroxy group (OH). An alternative view of an AHA explains the name: this sees the carboxylic acid as the key component; the carbon atom attached to it is then called the alpha carbon, and it is on this that the hydroxy group is located. (A beta-hydroxy acid (BHA) has the hydroxy group on the next but one carbon.)

The alpha carbon atom has two other atoms attached, and this is why there are many possible AHAs, although only a few are commercially important. The simplest AHA is glycolic acid, also known as hydroxy-acetic acid or hydroxyethanoic acid. Its formula is $HOCH_2CO_2H$, and the other two atoms attached to the alpha carbon are merely hydrogen atoms.

Lactic acid is also a relatively simple molecule, in that the alpha carbon has attached to it a hydrogen and a methyl group (CH_3). This introduces an extra dimension, however, in that two molecular structures are possible; these are mirror images of each other, i.e. a left-handed and a right-handed arrangement—see **chirality**. The two forms are known as D-lactic acid and L-lactic acid, where D stands for *dexter*, the Latin for 'right', and L stands for *laevus*, meaning 'left'. Industrially produced lactic acid is normally a mixture of the D and L forms, but in living things one form tends to dominate, and the human body produces mainly L-lactic acid, which is present in blood, muscle, liver, kidneys, and other organs.

Other natural AHAs are more complex in terms of the other two groups attached to the alpha carbon. Citric acid has two more acid groups, in the form of CH_2CO_2H, giving it three acid groupings in total. The less common malic acid has a CH_2CO_2H group and a hydrogen atom attached to its alpha carbon. Mandelic acid has a benzene group and a hydrogen atom attached to its alpha carbon. Tartaric acid is, in effect, two glycolic acids joined together at the alpha carbons.

Amino acids are the building blocks of proteins, and consist of a carbon atom to which is bonded an amine group (NH_2) and an acid group (CO_2H), with various other groups occupying the other two sites. The simplest amino acid of all, glycine ($NH_2CH_2CO_2H$), has two hydrogen atoms attached. Replace one of these hydrogens by a methyl group (CH_3) and you have alanine, and this is a molecule that can exist in two forms that are mirror images of each other. The one that occurs naturally is L-alanine—see also **chirality**.

There are around 200 types of amino acid found in Nature, and human protein has 22 of them. Ten of these are called **essential** amino acids because the body cannot make them, so they are an essential part of our diet.

Buffering an acid solution is a way of keeping its pH stable, and this is done by adding a salt of the acid concerned. For example, sodium hydrogen carbonate is added to carbonic acid to buffer it. A buffer ensures that, if something starts to neutralize that acid, more of it will form from the salt to compensate, thereby keeping the pH stable. Alkaline solutions can also be buffered by adding a salt of the base to a weak base: for example, adding ammonium chloride to ammonium hydroxide solution will keep its pH stable.

Buffers are particularly important in living organisms, which can be very sensitive to changes in pH. Many industrial processes, such as dyeing and fermentation, also require the use of buffers, and in the food industries buffers are added to stabilize acidity.

Chirality is the term used to describe the phenomenon of molecular structure in which two molecules have exactly the same number of atoms, arranged and bonded in exactly the same way, yet they are not the same because they are mirror images of each other. This situation will arise whenever there is a carbon atom in a molecule that forms bonds to four different atoms or groups. There is then a left- and a right-hand form of the molecule, more correctly described as a pair of **enantiomers**, or a chiral pair. The older term for this phenomenon was 'optical isomers' because of the different ways they interacted with light, and this was how they could be distinguished.

Nowadays, chemists use the symbols **R** and **S**, again based on right and left, in this case from the Latin words *rectus* and *sinister*, and they identify exactly which of the two forms of the molecules is being referred to. The assigning of **R** or **S** depends on the nature of the groups that are present, whereas the earlier terms of D (*dexter*, 'right') and L (*laevus*, 'left') were simply relative to each other, and you cannot assume D in the old system will be **R** in the new. Nevertheless, the older, long-established terms are still used in the life sciences and medical sciences.

If you wonder how two molecules with exactly the same atoms joined by exactly the same bonds can really be different, then you need look no further than your own hands for an answer. Each hand has a thumb, index finger, middle finger, ring finger, and little finger, and all are attached to the palm and all are arranged in the same order. Yet each hand is different, and we are reminded of that fact if we try to fit our right hand into a left-hand glove. Just as a right hand fits only a right glove, so a right molecule might fit only a right-shaped receptor.

Diagonal relationships between adjacent groups of the periodic table lead to chemical similarities that would not be expected. Thus, lithium is the first element of group 1, and it often behaves more like magnesium, which is the second element of group 2, than like sodium which is the second element of group 1. The same relationship applies to beryllium (the first element of group 2) and aluminium (the second element of group 3), to boron (the first element of group 3) and silicon (the second element of group 4), and so on, although the effect gradually disappears across the groups of the periodic table. The explanation for diagonal relationships is that the atoms in question are often similar in size, and size often determines how an atom will behave.

Double bonds are stronger, shorter, and more rigid than single bonds, but they are also more reactive. A single bond is formed between two atoms when they share two electrons. A double bond involves them sharing four electrons. (A triple bond has six electrons.) The rigidity of a double bond means that the atoms or groups attached to each end can be located on the same side of the double bond or on different sides. The former is referred to as the *cis* arrangement, and the latter is called the *trans* arrangement.

EDTA is short for ethylene diamine tetra acetic acid. This molecule consists of a core grouping of $N-CH_2-CH_2-N$, with two acetic acid groups ($-CH_2CO_2H$) attached to each nitrogen. Some of these four acid groups may be neutralized to make the EDTA more soluble.

Ester is the name given to the product formed from an acid and an alcohol. If the acid is RCO_2H (where R can be any organic group) and the alcohol is $R'OH$ (where again R' can be any organic group), then the ester formed from them would be RCO_2R'. (The other product of the reaction is H_2O.) Esters generally have a strong and attractive smell, and as such give rise to the aromas of many fruits.

Fatty acids are organic compounds consisting of a hydrocarbon chain with an acid group (CO_2H) at the end. To qualify as a fatty acid the chain must be more than just a few carbon atoms long, although some authors put even the short chains of one, two, or three carbons in this category. The best-known fatty acids are palmitic and stearic acids with 16 and 18 carbons, respectively (including the carbon of the acid group). The carbon chains may be bonded entirely with single bonds between the carbon atoms along the chain, and as such are called *saturated* fatty acids, or they may have one or more **double bonds** along the chain, in which case they are referred to as *mono-unsaturated* and *polyunsaturated* fatty acids, respectively. See also **triglycerides**.

The effects of molecular structure on physical properties can be seen by comparing the melting points of the three types of 18-carbon fatty acids: stearic acid, which is saturated, melts at 72 °C, while elaidic acid, which has a *trans* **double bond** half-way along its chain, melts at 44 °C, and oleic acid, which has a *cis* double bond half-way along its chain, is an oil at room temperature, and when solidified it melts at 13 °C.

Free radical. This is a molecule that has a single unbonded electron. As such, it is highly reactive because this electron gives it the power to attack almost anything. Use can be made of free radicals—to initiate polymerization reactions, for example—but free radicals are generally to be avoided in living things because of the harm they can do.

Free radicals rapidly attack carbon–hydrogen bonds, the most common of all bonds, which are present in almost all organic molecules, and they abstract the hydrogen. Free radicals also add to double bonds. Sometimes, two free radicals will combine with each other, thereby pairing off the free electrons and so forming a stable molecule.

The human body wages an endless war against free radicals, which are formed as we use the oxygen that we breathe in. To protect against free-radical damage, the body uses vitamins C and E; the former is soluble in water and the latter is soluble in fats, and together they are able to deactivate almost any free radical that they come across in virtually every part of the body.

Nevertheless, there are some relatively stable free radicals, such as nitric oxide (NO), which, like most free radicals has one free electron, and oxygen (O_2), which has *two* such electrons, one on each oxygen atom. Both molecules play an important part in living things: see Chapter 3 for the NO story.

Free radical polymerization. To get double-bonded molecules to polymerize, it is necessary to add an electron to them, and this is done by bringing them into contact with a free radical, or a molecule that easily generates such free radicals. The free radical passes its free electron to the **double bond** of a monomer, thereby turning that molecule into a free radical, which can then react with another monomer molecule in a way that links them together; but this still leaves the product with the free electron, which can then attack another monomer, and so on. The simplest molecule to undergo such a reaction is ethylene (official name, ethene), and the product is polyethylene (also known as polyethene or polythene). In chemical terms, the process is

$$CH_2{=}CH_2 + CH_2{=}CH_2 \rightarrow -CH_2{-}CH_2{-}CH_2{-}CH_2{-} \rightarrow (-CH_2)_n{-}$$

where n can run into millions.

Once the polymerization has been initiated, the polymers continue to grow until they either lose the free electron or combine with another free radical, when the two electrons will unite to form an unreactive electron pair. The extent of polymerization can be controlled by the amount of free-radical initiator used; the less of this there is, the longer the polymer chains are likely to be.

Glycerin, glycerine, glycerol are all names for the same simple chemical, which is common in Nature as well as being an important resource of the chemical industry, and a by-product of soap manufacture and fatty-acid production. Global production of glycerol already exceeds 6 million tonnes a year, and it is used in foods, cosmetics, and lubricants, while industry converts it to polymers, plasticizers, and explosives. Its chemical formula is $C_3H_8O_3$ and it is, in effect, a triple alcohol with a chain of three carbon atoms, each of which has attached to it an alcohol group (OH), so that its molecular formula is better written as $CH_2(OH)CH(OH)CH_2OH$; its systematic chemical name is propan-1,2,3-triol. Fats and oils are derivatives of glycerol, known as **triglycerides.**

Hydrocarbon polymers. The monomers from which rubbery hydrocarbons are made are isobutylene, butadiene, isoprene, and styrene. Each of these simple molecules has the necessary double bond that enables it to couple to form polymers.

Isobutylene has two methyl groups (CH_3) on one of the carbon atoms of a double bond, and the product of its polymerization is PBI (short for polyisobutylene). Butadiene has two double bonds with hydrogen atoms attached, and is written $CH_2=CH–CH=CH_2$. When these molecules link up, only one double bond is needed for the process, which means the polymer that forms still has double bonds along its backbone:

$$CH_2=CH–CH=CH_2 + CH_2=CH–CH=CH_2 \rightarrow$$
$$–CH_2–CH= CH–CH_2–CH_2–CH=CH–CH_2–$$

Isoprene also has two double bonds; in fact, it is butadiene with a methyl group attached to one of the middle carbons, and its correct chemical name is 2-methyl-but-1,3-diene. When this polymerizes it, too, forms a polymer that retains one of the double bonds. Things are a little more complicated than this, in that isoprene can give two types of polymer, called *cis*-polyisoprene and *trans*-polyisoprene depending on the polymerization process and defined by the way the chain is arranged at the site of the double bond.

The *cis* form is obtained when sap from rubber trees polymerizes, and it is springy and soft, whereas the *trans* variety is obtained from the sap of the tree *Palaquium oblongifolia*, which gives gutta-percha; this is much harder because the *trans* polymer molecules can pack together more tightly than *cis* polymer molecules.

Styrene has a benzene ring attached to one of the carbons of the ethylene double bond, and its polymer, polystyrene, can be used to make all kinds of plastic materials. When styrene and butadiene are polymerized together, however, the resulting co-polymer, SBR (short for styrene butadiene rubber), is rubbery and finds widespread use—see Chapter 6.

Hydrogen bonding is a weak form of chemical interaction, but it nevertheless plays a pivotal role in all living things, even determining the structure of proteins and of DNA itself. Hydrogen bonding explains why water is a liquid and why ice floats. The most important type of hydrogen bond is between two oxygen atoms, one of which must have a hydrogen atom attached. This hydrogen is then attracted to another oxygen and a hydrogen bond forms, and is depicted in chemistry as O–H⋯O, the dotted line indicating that the second bond is longer and weaker. Hydrogen bonds can also form between nitrogen atoms and between an oxygen and a nitrogen atom, and these are the ones that hold together the strands of DNA. Although hydrogen bonds are weak compared with normal bonds, having less than a quarter of the energy of these, there are many of them and quantity makes up for quality.

Water is the best example of the power of hydrogen bonding. H_2O is a very light molecule, and might be expected to be a gas at room temperature, as is hydrogen sulfide (H_2S), which is twice its weight. Yet, water is a liquid and does not boil until it reaches the relatively high temperature of 100 °C (H_2S boils at *minus* 60 °C). In water the H_2O molecules form three hydrogen bonds to surrounding water molecules, whereas in ice they form four, and this results in a more open lattice structure which is actually less dense than the liquid. As a result, when water freezes it expands, instead of contracting like most other liquids, and this means that ice floats instead of sinking.

Nano, as in nanotechnology and nanoscale, is based on the unit of length, the nanometre, which is a billionth of a metre, i.e. 10^{-9} metre. Atoms generally have a radius of between 0.1 and 0.2 nanometres, and the world of molecules is one of nanoscale dimensions. It is now possible to work on things of this size, and research is going on in chemistry laboratories all round the world to make nanosized devices, the ultimate object being to create nano-machines and even nano-computers.

Natural oils consist of **fatty acids**, fatty **esters**, or both. A commonly used oil is castor oil, pressed from the seeds of the castor plant *Ricinus communis*. In this case, the main component is a rather unusual fatty acid, ricinoleic acid, which has a chain of eighteen carbon atoms with a hydroxy group on the twelfth carbon. It also has a double bond between carbons 9 and 10. Esters of ricinoleic acid are used in all kinds of cosmetics, and the acid itself can also be modified by making changes to the hydroxy group. Such is the versatility of these derivatives that they are used in moisturizers, antiperspirants, hair oils, colourants, perfumes, and even nail varnishes (where they impart a degree of suppleness to the dried varnish to prevent it from cracking easily).

Lanolin is a mixture of esters based on thirty-six fatty acids combined with thirty-three long-chain alcohols.

Osmosis. This describes the passage of a solvent (generally water) through a semi-permeable membrane which is separating two solutions of different concentration. The membrane needs to have pores small enough to permit solvent molecules to pass through but not the things that are dissolved in it. The solvent will diffuse from the less concentrated side to the more concentrated side, so as to make the concentrations on both sides of the membrane equal, at which point osmosis ceases. If pressure is applied at the more concentrated side then osmosis can be prevented, and this pressure is called the osmotic pressure. If an even greater pressure is applied, then reverse osmosis will occur, with the solvent moving out of the more concentrated solution. This method is used in some desalination plants to produce fresh water from seawater, and it requires pressures in excess of 25 atmospheres.

pH is the term used to describe acidity and alkalinity; it is really a measure of the concentration of the molecular ion species H_3O^+ (often just written as H^+), which is the active component formed by acids when they react with water. The concentration of this ion can vary so widely that a special scale, known as pH, is needed to measure it. The pH is a logarithmic scale in which this concentration is expressed in terms of negative powers of ten. For example, in neutral water the concentration of H_3O^+ is very low, only 10^{-7} ions per litre, and its pH is defined as 7. The pH scale is an inverse scale as far as acids are concerned: in other words, the lower the pH number, the stronger the acid. The range of normal acidities is from 1 (the strongest) to 7 (neutral water) and, being logarithmic, means that an acid of pH 1 has a million times more H_3O^+ than neutral water of pH 7.

The pH scale can also be extended to cover alkaline conditions from 7 to 14, where OH^- predominates and the concentration of H_3O^+ falls by a further factor of a million. The full span of pH from 1 to 14 represents a difference in acidity of a *million million* times. Nevertheless, we can come across materials in our everyday life which encompass the full range:

Commonly encountered acids and alkalis

---

pH	Typical substances	Acid or alkali responsible
0*	chemical reagent	concentrated sulfuric acid
1	stomach acid, battery acid	dilute hydrochloric acid, and dilute sulfuric acid, respectively
2	lemon juice	citric acid
3	vinegar	acetic acid
4	tomato juice	ascorbic acid (vitamin C)
5	beer, rainwater	carbonic acid (H_2CO_3)
6	milk	lactic acid
7	blood	neutral conditions
8	seawater	dissolved calcium carbonate
9	bicarbonate solution	sodium hydrogen carbonate in water
10	milk of magnesia	magnesium hydroxide in water
11	household ammonia	ammonia solution in water
12	garden lime	calcium hydroxide powder
13	drain cleaner	sodium hydroxide solution
14	caustic soda	concentrated sodium hydroxide solution

* There are some very strong acids with negative pH values, but these are not met with in everyday life.

Polymers are molecules in which there is a long chain of atoms with the same repeating unit. The simplest is polyethylene, in which the chain consists only of carbon atoms with hydrogens attached; in other words it is $-CH_2-CH_2-CH_2-$, and so on *ad infinitum*, or at least a million times or more. As the name polyethylene implies, this is made up of many ethylenes (the prefix poly is derived from the Greek *polloi* meaning 'many'); ethylene is the basic molecule from which it is constructed and is called the monomer (mono comes from the Greek *monos* meaning 'alone'). Ethylene is more properly called ethene; it has the chemical formula $CH_2=CH_2$, and its **double bond** can be traded for two single bonds to other carbon atoms, so linking it to other ethylenes in a polymerization process that produces the long chains.

Polymers can be made from all kinds of monomers, some being simply derivatives of ethylene with another atom or group of atoms replacing one or more of its hydrogen atoms. For example, if the ethylene molecule has a chlorine atom attached, the polymer product is PVC, polyvinyl chloride. This name comes from the monomer $CH_2=CHCl$, which is known in industry as vinyl chloride, the word vinyl being the old name for the $CH_2=CH$ grouping of atoms. If this polymer were discovered today, it would almost certainly be called PCE, short for polychloroethene.

Polyacrylic acid, which is mentioned in Chapter 6, is made from the monomer of formula $CH_2=CH-CO_2H$, but if some of the acid has been neutralized to the sodium salt $CH_2=CH-CO_2Na$, then the product is called polyacrylate. The mixture is polymerized by what is known as a **free-radical** mechanism, and this requires the addition of a **free-radical** initiator. The initiator generally used to polymerize acrylic acid is persulfate ions, the action of which is triggered by heating, or by adding sodium thiosulfate.

Quaternary ammonium compounds are derivatives of the simple ammonium ion, NH_4^+. This is an ammonia (NH_3) molecule that has picked up a hydrogen ion H^+, something that it is inclined to do whenever it encounters an acid. Between one and four of the hydrogens can be replaced by organic components bonded to the nitrogen through a carbon atom, and these molecules are known as quaternary ammonium compounds. There are hundreds of simple organic groups that can be so attached; in theory, therefore, there are millions of possible quats although only a few are commercially important (see Chapter 5).

Triglyceride is the correct name for fats and oils; they are all derivatives of **glycerol**, in which a **fatty acid** group is attached to each of the three oxygens on the three carbon atoms of this molecule, thereby forming **esters** (see above). The fatty acids can be saturated, unsaturated, polyunsaturated, etc. (see Chapter 2 for more details), and in an individual triglyceride they can all be the same or all different. Consequently, there are many possible triglycerides because there are so many ways that three fatty acids can be selected from all of those that are known. For example, milk contains more than 150 chemically distinguishable triglycerides. Plants and animals have generally evolved to make only a few triglycerides.

Some of these triglyceride molecules are *isomers* of one another, by which is meant that they differ only in the order in which the fatty acids are attached to the three atoms of glycerol, C_1–C_2–C_3. If there are three different fatty acids attached to the glycerol then there are three possible versions of that molecule, depending on which fatty acid is attached to the middle carbon.

Units in chemistry are concerned with the molecular world and hence describe small amounts, and these can be related to everyday language. Thus, *milli* is a prefix that indicates a thousandth part (numerically expressed as 10^{-3}); *micro* refers to a millionth (10^{-6}); *nano* to a billionth (10^{-9}); and *pica* to a trillionth (10^{-12}). These quantities are hard to visualize, but in terms of weight a milligram is about the weight of a grain of sand, and a microgram the weight of a speck of dust. A nanogram would be invisible to the naked human eye.

Because atoms weigh so little, a defined unit of weight has been chosen. This is the *mole*, which is defined as the number of atoms in 12 grams of carbon. Numerically, this comes to 6.022×10^{23} atoms, sometimes referred to as the Avogadro number. The atomic weight of an element is the weight in grams of this number of its atoms, and the molecular weight is the total of all the atomic weights of all the atoms of which a molecule is composed. Such a quantity is often too large to be of use when it comes to dilute solutions, and then the quantities used are *millimoles* (one thousandth of a mole) per litre, or even *micromoles* (one millionth of a mole) per litre.

Waxes are predominantly hydrocarbons; in other words, they consist of molecules with chains of carbon atoms to which are attached only hydrogen atoms, although they may have an acid group (CO_2H) or an alcohol group (OH) at the end of the chain. Hydrocarbon chains are composed of CH_2

groups linked together, and as they get longer so the properties change from gases, to liquids, to waxy solids. They are used in ointments, polishes, and candles.

Plants and animals produce waxes because hydrocarbons are water repellent, and they offer protection against dehydration and microbial attack. Nature finds it more convenient to make waxes by linking hydrocarbon chains, and these are formed from long-chain **fatty acids** with between 18 and 28 carbon atoms in the chain. A typical one found in waxes is cretoic acid, which consists of a hydrocarbon chain of 25 carbon atoms with an acid group (CO_2H) at the end, i.e. $C_{25}H_{51}CO_2H$. Although the more common dietary fatty acids, such as palmitic acid $C_{16}H_{33}CO_2H$, are liquids, they, too, can be the basis of a wax when a second long hydrocarbon chain is attached to the acid group to form an **ester**.

Some waxes are long-chain alcohols; for example carnaubyl alcohol has 24 carbon atoms in its chain and the alcohol OH group attached at one end, i.e. $C_{24}H_{49}OH$. The long-chain alcohols with shorter chains are liquids, such as stearyl alcohol, $C_{17}H_{35}OH$, and hexadecanol, $C_{16}H_{33}OH$, but again this does not exclude them from being waxy if they are part of an ester. Indeed, most natural waxes are esters; for example, the myricin wax of beeswax is an ester of palmitic acid and has the formula $C_{15}H_{31}CO_2C_{31}H_{63}$.

Further Reading

General reading

Atkins, P., *The Periodic Kingdom*, London, Weidenfeld & Nicolson, 1995.

Ball, P., *Stories of the Invisible*, Oxford, Oxford University Press, 2001.

Ball, P., *Bright Earth*, London, Viking, 2001.

Ball, P., *The Ingredients*, Oxford, Oxford University Press, 2002.

Bowen, H. J. M., *Environmental Chemistry of the Elements*, London, Academic Press, 1979.

Brock, W. H., *The Fontana History of Chemistry*, London, Fontana, 1992.

Büchner, W., Schliebs, R., Winter, G., and Büchel, K. H., *Industrial Inorganic Chemistry*, Weinheim, Germany, VCH, 1989.

Budavari, S. (ed.), *The Merck Index*, 11th edn, Rahway, NJ, Merck & Co., 1989.

Daintith, J. and Gjertsen, D. (eds), *Oxford Dictionary of Scientists*, Oxford, Oxford University Press, 1999.

Diem, K. and Lentner, C. (eds), *Scientific Tables*, 7th edn, Basel, Switzerland, Documenta Giegy, 1970.

Emsley, J., *Nature's Building Blocks*, Oxford, Oxford University Press, 2001.

Emsley, J. and Fell, P., *Was It Something You Ate?*, Oxford, Oxford University Press, 1999.

Gray, H. B., Simon, J. D., and Trogler, W. C., *Braving the Elements*, Sausalito, Calif., University Science Books, 1995.

Hunter, D., *The Diseases of Occupations*, 5th edn, London, Hodder and Stoughton, 1976.

Kutsky, R. J., *Handbook of Vitamins, Minerals and Hormones*, 2nd edn, New York, Van Nostrand Reinhold, 1981.

Lane, N., *Oxygen*, Oxford, Oxford University Press, 2002.

Le Couteur, P. and Burreson, J., *Napoleon's Buttons*, New York, Penguin Putnam, 2003.

Marshall, P., *The Philosopher's Stone*, London, Macmillan, 2001.

Muir, H. (ed.), *Larousse Dictionary of Scientists*, Edinburgh, Larousse, 1994.

Quadbeck-Seeger, H.-J. (ed.), *World Records in Chemistry*, Weinheim, Germany, Wiley-VCH, 1999.

Sax, N. I. and Lewis Sr, R. J., *Dangerous Properties of Industrial Materials*, 7th edn, New York, Van Nostrand Reinhold, 1989.

Selinger, B., *Chemistry in the Market Place*, 5th edn, Sydney, Harcourt Brace Jovanovich, 1998.

Selinger, B., *Why the Watermelon Won't Ripen in Your Armpit*, St Leonards, NSW, Allen & Unwin, 2000.

Snyder, C. H., *The Extraordinary Chemistry of Ordinary Things*, New York, John Wiley & Sons, 1992.

Strathern, P., *Mendeleyev's Dream: The Quest for the Elements*, London, Hamish Hamilton, 2000. ˀ

Wade, A. (ed.), *Martindale, the Extra Pharmacopoeia*, London, The Pharmaceutical Press, 1977.

Weeks, M. E. and Leicester, H. M., *Discovery of the Elements*, 7th edn, Easton, PA, Journal of the American Chemical Society, 1968.

Williams, J. P. and Fraústo da Silva, J. J. R., *The Biological Chemistry of the Elements*, Oxford, Clarendon Press, 1991.

Williams, J. P. and Fraústo da Silva, J. J. R., *The Natural Selection of the Chemical Elements*, Oxford, Clarendon Press, 1996.

Books and articles providing more in-depth coverage of specific items covered in the various chapters:

Chapter 1

Cotton, F. A. and Wilkinson, G., *Advanced Inorganic Chemistry*, 5th edn, New York, Wiley Interscience, 1988. [Boron nitride]

Kimbrough, D. R., 'The photochemistry of sunscreens', *Journal of Chemical Education*, January 1997.

Pallingston, J., *Lipstick*, London, Simon & Schuster, 1999.

Raber, L., 'Self Tanners', *Chemical and Engineering News*, 12 June 2000.

Ragas, M. C. and Kozlowski, R., *Read My Lips: A Cultural History of Lipstick*, San Francisco, Chronicle Books, 1998.

Reisch, M., 'Sunscreens', *Chemical and Engineering News*, 24 June 2002.

Roscoe, H. E. and Schorlemmer, C., *A Treatise on Chemistry*, London, Macmillan, 1920. [Boron nitride]

Woodruff, J., *Manufacturing Chemist*, May 1995; May 1997; July 1998; June 1999; July 2001.

Young, A., 'Sunlight, the skin and sunscreens', *The Biochemist*, Feb./March 1994.

Chapter 2

Addiscott, T. M., Whitmore, A. P., and Powlson, D. S., *Farming, Fertilizers and the Nitrate Problem*, Wallingford, Oxon, CAB International, 1991.

Al-Dabbagh, S., Forman, D., Bryson, D., Stratton, I., and Doll, E. C., 'Mortality of nitrate fertilizer workers', *British Journal of Industrial Medicine*, 1986, vol. 43, p. 507.

Belitz, H.-D. and Grosch, W., *Food Chemistry*, Heidelberg, Germany, Springer-Verlag, 1987.

Carpenter, K., *The History of Scurvy and Vitamin C*, Cambridge, Cambridge University Press, 1986.

Coultate, T., *Food: The Chemistry of its Components*, 4th edn, Cambridge, Royal Society of Chemistry, 2002.

Davies, M. B., Austin, J., and Partridge, D. A., *Vitamin C: Its Chemistry and Biochemistry*, London, Royal Society of Chemistry, 1991.

Dietary Reference Values for Food Energy and Nutrients for the UK, London, Dept of Health, HMSO, 1991.

Drummond, J. C. and Wilbraham, A., *The Englishman's Food*, London, Cape, 1939; revised by D. Hollingworth and published by Pimlico, London, 1994. [An excellent account of scurvy down the ages.]

Fatty Acids: 7th Supplement to McCance & Widdowson's The Composition of Foods, Cambridge, compiled by the UK Ministry of Agriculture, Fisheries and Food, Royal Society of Chemistry, 1998.

Forman, D., Al-Dabbagh, S., and Doll, E. C., 'Nitrates, nitrites and gastric cancer in Great Britain', *Nature*, 1985, vol. 313, p. 620.

Goertzel, T. and Goertzel, B., *Linus Pauling*, New York, Basic Books, 1996.

L'Hirondel, J.-L., 'Dietary nitrates pose no threat to human health', *Environmental Health*, Chapter 6, London, Butterworth-Heinemann, 1999.

L'Hirondel, J. and L'Hirondel, J.-L., *Nitrate and Man: Toxic, Harmless, or Beneficial?*, Wallingford, Oxon, CABI Publishing, 2002.

Martin, C., 'TFAs—a fat lot of good?', *Chemistry in Britain*, October 1996, p. 34.

Sanders, T., *Dietary Fats*, London, a 54-page briefing paper produced for the UK Health Education Authority, 1994.

Smil, V., *Enriching the Earth: Fritz Haber, Carl Bosch and the Transformation of World Food*, Boston, Mass., MIT Press, 2001.

Wilson, W. S., Ball, A. S., and Hinton, R. H., (eds), *Managing Risks of Nitrates to Humans and the Environment*, Cambridge, Royal Society of Chemistry, 1999.

Chapter 3

Ainscough, E. W. and Brodie, A. M., 'Nitric oxide—some old and new perspectives', *Journal of Chemical Education*, August 1995, p. 686.

Combs, G. F. Jr and Combs, S. B., *The Role of Selenium in Nutrition*, Orlando, Fla., Academic Press, 1986.

Djerassi, C., *NO*, Athens, Georgia, University of Georgia Press, 1998.

Harlow, G. (ed.), *The Nature of Diamonds*, Cambridge, Cambridge University Press, 1998.

Lewis, A., *Selenium: The Facts About This Essential Mineral*, Wellingborough, Thorsons, 1982.

Lewis, A., *Selenium, The Essential Element You Might Not be Getting Enough of*, Wellingborough, UK, and Rochester, USA, Thorsons, 1982.

Passwater, R., *Selenium as Food & Medicine*, New Canaan, Conn., Keats Publishing, 1980.

Rayman, M. P., 'The argument for increasing selenium intake', in *Proceedings of the Nutrition Society*, 2002, vol. 61, p. 203.

Shamberger, R. J., *Biochemistry of Selenium*, New York, Plenum, 1983.

Stone, T. and Darlington, G., *Pills, Potions, and Poisons: How Drugs Work*, Oxford, Oxford University Press, 2000.

Chapter 4

Anon., *Dettol: Protection from Infection*, company product monograph, Hull, UK, Reckitt & Colman, 1996.

Anon. *Understanding Germs, Hygiene & Health*, produced by the Catalyst Science Communication and Consultancy Ltd, and the Chemical Industries Education Centre at the University of York, 1999.

Thornton, J., *Pandora's Poison: Chlorine, Health and a New Environmental Strategy*, Boston, Mass., MIT Press, 2000.

Zündorf, U., *Chemistry With Chlorine: Opportunities, Risks, Perspectives*, Leverkusen, Germany, Bayer AG Publications, 1995—versions in German and English. [This is one of several booklets produced by companies for the general public, and one of the best.]

Chapter 5

Aldridge, S., *Magic Molecules*, Cambridge, Cambridge University Press, 1998.

Altman, P., Cunningham, J., Dhanesha,U., *et al.*, 'Disturbance of cerebral function in people exposed to drinking water contaminated with aluminium: retrospective study of Camelford water incident', *British Medical Journal*, 1999, vol. 319, p. 807.

Barondes, S. H., *Better than Prozac*, Oxford, Oxford University Press, 2003.

Graske, A., Thuvander, A., Johannisoon, A. *et al.*, 'Influence of aluminium on the immune system: an experimental study on volunteers', *Biometals*, 2000, vol. 13, p. 123.

Hall, M. *Target Depression*, London, Association of British Phamaceutical Industries' booklet, 1999.

Henry, J. (ed.), *BMA Guide to Medicines & Drugs*, London, Dorling Kindersley, 1993.

Hughes, J. T., *Aluminium and Your Health*, Cirencester, Glos., Rimes House Publishing, 1992.

Johnson, F. N. (ed.), *Depression and Mania: Modern Lithium Therapy*, Oxford, IRL Press, 1987.

Landsberg, J. P., McDonald, B., and Watts, F., 'Absence of aluminium in neuritic plaque cores in Alzheimer's disease, *Nature*, 1992, vol. 360, p. 65.

Makjanic, J. and Watt, F., 'Nuclear microscopy in Alzheimer's disease', *Nuclear Instruments and Methods in Physics Research B*, 1999, vol. 150, p. 167.

Maletzky, B. and Blachly, P. H., *The Use of Lithium in Psychiatry*, London, Butterworths, 1971.

Schrauzer G. N. and Klippel, K. F. (eds), *Lithium in Biology and Medicine*, Weinheim, Germany, Wiley-VCH, 1991.

Timbrell, J. A., *Introduction to Toxicology*, London, Taylor & Francis, 1989.

Twort, A. C., Law, F. M., Crowley, F. W., and Ratnayaka, D. D., 'Storage, sedimentation, clarification and filtration', Chapter 7 of *Water Supply*, 4th edn, London, Edward Arnold, 1994.

Chapter 6

Colborn, T., Dumanoski, D., and Myers, J., *Our Stolen Future*, New York, Dutton, Penguin Books, 1996.

Dimond, S. S., Waechter, J. M. Jr., Breslin, W.J., *et al.*,'Evaluation of reproductive organ development in the male offspring of female wistar rats exposed to bisphenol A in the drinking water', and 'Evaluation of reproducing organ development in CF-1 mice after prenatal exposure to bisphenol A'.

These papers can be accessed via the Internet on:

<www.bisphenol-a.org/evaluation_of_reproductive_organ.htm>
<www.bisphenol-a.org/evaluation_of_reproductive _organ2.htm>

Edgar, W. M. and O'Mullane, D. M. (eds), *Saliva and Health*, 2nd edn, London, British Dental Association, 1999.

Howdeshell, K.L., Hotchkiss, A.K., Vom Saal, F. S. *et al.*, 'Exposure to bisphenol A advances puberty', *Nature*, 1999, vol. 401, p. 763. [Other papers by Vom Saal can be accessed via <http://www.missouri.edu>.]

Lobos, J. H., Leib, T. K., and Su, T. M., 'Biodegradation of bisphenol A and other bisphenols by a gram-negative aerobic bacteria', *Applied and Environmental Microbiology*, 1992, vol. 58. p. 1823.

Pulgar, R., Olea-Serrano, M.F., Novillo-Fertrell, A., *et al.*, 'Determination of bis-phenol A and related aromatic compounds released from bis-GMA-based composites and sealants by HPLC', *Environmental Health Perspectives*, 2000, vol. 108, p.21. [This contains sixty-three references, many of them to studies of BPA.]

Read, J., 'On the road', *Chemistry in Britain*, August 1998. [Also discusses orimulsion.]

Voirol, F., *The Evolution of Chewing Gum*, Basel, Switzerland, Xyrofin Ltd, 1985. [This was a booklet designed to promote the alternative sweetener xylitol, but it is a useful source of data about chewing gum, and its ingredients and manufacture.]

Postscript

Allaby, M., *Facing the Future*, London, Bloomsbury, 1995.

Bate, R. (ed.), *What risk? Science, Politics and Public Health*, Oxford, Butterworth-Heinemann, 1997.

Henderson, M., *Living with Risk*, The British Medical Association, and John Wiley & Sons, Chichester, UK, 1987.

Michaelis, A.R., 'Stop: Chemiphobia', *Interdisciplinary Science Reviews*, 1996, vol. 21, p. 130.

Milloy, S. J., *Junk Science Judo*, Washington, DC, Cato Institute, 2001.

Paulos, J. A., *A Mathematician Reads the Newspapers*, London, Penguin, 1996.

Index